Cold War Science and the Transatlantic Circulation of Knowledge

History of Science and Medicine Library

VOLUME 51

History of Modern Science

Editors

Kostas Gavroglu (*Athens University*)
Massimiliano Badino (*Universitat Autònoma de Barcelona*)
Jürgen Renn (*Max Planck Institute for the History of Science, Berlin*)

VOLUME 1

The titles published in this series are listed at *brill.com/hims*

Cold War Science and the Transatlantic Circulation of Knowledge

Edited by

Jeroen van Dongen

Associate editors

Friso Hoeneveld
Abel Streefland

BRILL

LEIDEN | BOSTON

Cover illustration: A 1971 file photo of a French thermo nuclear bomb detonated at the Fangataufa atoll, French Polynesia.
© AP/Hollandse Hoogte.

Library of Congress Cataloging-in-Publication Data

Names: Dongen, Jeroen van, 1974-
Title: Cold War science and the transatlantic circulation of knowledge / edited by Jeroen van Dongen.
Description: Leiden : Brill, 2015. | Series: History of science and medicine library, ISSN 1872-0684 ; volume 51 | Series: History of modern science ; volume 1 | Includes bibliographical references and index.
Identifiers: LCCN 2015036416| ISBN 9789004264212 (hardback : acid-free paper) | ISBN 9789004264229 (e-book)
Subjects: LCSH: Science and state—United States—History—20th century. | Science and state—Europe, Western—History—20th century. | United States—Relations—Europe, Western. | Europe, Western—Relations—United States. | Knowledge, Sociology of—History—20th century. | Cold War—Social aspects.
Classification: LCC Q127.U6 C623 2015 | DDC 338.9492/0609045—dc23 LC record available at http://lccn.loc.gov/2015036416

This publication has been typeset in the multilingual "Brill" typeface. With over 5,100 characters covering Latin, IPA, Greek, and Cyrillic, this typeface is especially suitable for use in the humanities. For more information, please see www.brill.com/brill-typeface.

ISSN 1872-0684
ISBN 978-90-04-26421-2 (hardback)
ISBN 978-90-04-26422-9 (e-book)

This book is printed on acid-free paper.

Contents

List of Illustrations and Tables VII
Note on Contributors VIII

Introduction 1

PART 1
Secrecy and Science

1 Scientists, Secrecy, and Scientific Intelligence: The Challenges of International Science in Cold War America 11
Ronald E. Doel

2 A 'Need-To-Know-More' Criterion? Science and Information Security at NATO during the Cold War 36
Simone Turchetti

3 A Transnational Approach to US Nuclear Weapons Relationships with Britain and France in the 60s and 70s 59
John Krige

PART 2
Dutch Perspectives

4 Putting a Lid on the Gas Centrifuge: Classification of the Dutch Ultracentrifuge Project, 1960–1961 77
Abel Streefland

5 *Quid Pro Quo*: Dutch Defense Research during the Early Cold War 101
Jeroen van Dongen and Friso Hoeneveld

6 Chemical Warfare Research in the Netherlands 122
Herman Roozenbeek

7 **The Fulbright Program in the Netherlands: An Example of Science Diplomacy** 136
Giles Scott-Smith

PART 3
'Cold War' Science?

8 **The Absence of the East: International Influences on Science Policy in Western Europe during the Cold War** 165
David Baneke

9 **Colonial Crossings: Social Science, Social Knowledge, and American Power from the Nineteenth Century to the Cold War** 184
Jessica Wang

PART 4
Scientific Hubris

10 **Cold War Atmospheric Sciences in the United States: From Modeling to Control** 217
Kristine C. Harper

11 **Small State versus Superpower: Science and Geopolitics in Greenland in the Early Cold War** 243
Matthias Heymann, Henry Nielsen, Kristian Hvidtfelt Nielsen and Henrik Knudsen

12 **The Ford Foundation and the Measurement of Values** 272
Paul Erickson

Index of Names 289

List of Illustrations and Tables

Illustrations

5.1 G.J. Sizoo 104
5.2 SHAPE Technical Center in 1969 112
7.1 Distribution of Dutch Fulbright grantees among different disciplines in the years 1949–1960 148
7.2 Destination of Dutch Fulbright scholars for various periods 152
7.3 Distribution of Dutch Fulbright grantees to top-level US universities according to subject area, 1949–1960 155
11.1 The end of 'Project Crested Ice' 245
11.2 War time bases and weather stations in Greenland in 1945 255
11.3 Transport of a nuclear reactor into the tunnel system of Camp Century 261

Tables

7.1 Dutch physicists who received Fulbright grants to conduct research in the United States, 1949–1960 157
7.2 Dutch electrical, aerodynamic, hydrological, and chemical engineers who received Fulbright grants to conduct research in the United States, 1949–1955 159

Note on Contributors

David Baneke
(PhD Utrecht University, 2008) is Lecturer in History and Philosophy of Science at Utrecht University. His interests include the role of scientists in society, and the political, social and cultural aspects of modern science. Baneke's latest project concerns the history of the Dutch astronomical community in the twentieth century.

Ronald E. Doel
(PhD Princeton University, 1990) is Associate Professor of History at Florida State University. He writes on scientific internationalism in the twentieth century, the physical environmental sciences, and photographs as historical evidence. Doel recently served as Project Leader for the nine member, seven nation project "Colony, Empire, Environment: A Comparative International History of Twentieth Century Arctic Science," funded through the BOREAS initiative of the European Science Foundation.

Jeroen van Dongen
(PhD University of Amsterdam, 2002) is Professor of History of Science at the University of Amsterdam. He also teaches at Utrecht University. Van Dongen is the author of *Einstein's Unification* (Cambridge University Press, 2010).

Paul Erickson
(PhD University of Wisconsin–Madison, 2006) is Assistant Professor of History and of Science in Society at Wesleyan University. He co-authored *How Reason Almost Lost its Mind: The Strange Career of Cold War Rationality* (University of Chicago Press, 2013).

Kristine C. Harper
(PhD Oregon State University, 2003) is Associate Professor of History at Florida State University. She specializes in the history of the atmospheric and water-related earth sciences, particularly during the Cold War.

Matthias Heymann
(PhD Technical University of Munich, 1995) is Associate Professor of the History of Technology at the Centre for Science Studies, Aarhus University, Denmark. His research focuses on the history of environmental science and

technology. He currently leads the project "Shaping Cultures of Prediction: Knowledge, Authority, and the Construction of Climate Change."

Friso Hoeneveld

studied history at the University of Amsterdam and is working on a PhD thesis on Dutch physics in the early Cold War at Utrecht University. His research interests include twentieth-century science in the context of European-American relations.

Henrik Knudsen

(PhD Aarhus University, 2005) is Archivist and Senior Researcher at the Danish National Archives. Knudsen's research interests cover science and technology in the global Cold War, nuclear history, science policy, and diplomacy.

John Krige

(PhD University of Sussex, 1978) is the Kranzberg Professor in the School of History, Technology and Society, Georgia Tech, Atlanta, GA. He is a historian of science and technology who works at the intersection of that field with security studies and the history of US foreign relations in the Cold War. He authored *American Hegemony and the Postwar Reconstruction of Science in Europe* (MIT Press, 2006) and co-edited with Naomi Oreskes, *Science and Technology in the Global Cold War* (MIT Press, 2014).

Henry Nielsen

(PhD Aarhus University, 1973) is Professor Emeritus at Aarhus University's Centre for Science Studies. His research interests cover many aspects of the scientific and technological development of Denmark after 1900, and he has published extensively on Danish science in the context of the Cold War.

Kristian H. Nielsen

(PhD Aarhus University, 2001) is Associate Professor of History of Science and Science Communication at Aarhus University. Nielsen has published in journals such as *Annals of Science, British Journal for the History of Science, Centaurus, Historical Studies of the Natural Sciences, History of Technology, Public Understanding of Science, Science as Culture*, and *Science Communication*.

Herman Roozenbeek

(MA) is Senior Researcher at the Netherlands Institute of Military History. He studied history at Leiden University and has written extensively on the history of the Royal Netherlands Army during the Cold War.

Giles Scott-Smith
(PhD Lancaster University, 1998) holds the Ernst van der Beugel Chair in the Diplomatic History of Atlantic Cooperation since WW II at Leiden University. His most recent work, together with Stephanie Roulin and Luc van Dongen, is *Transnational Anti-Communism and the Cold War* (Palgrave, 2014).

Abel Streefland
(MSc Utrecht University, 2010) is preparing a PhD at Leiden University, where he studies Jacob Kistemaker's role in the development of uranium enrichment in the Netherlands.

Simone Turchetti
(PhD University of Manchester, 2003) is Lecturer at the Centre for the History of Science, Technology and Medicine at the University of Manchester. He has just completed the five-year project "The Earth Under Surveillance," funded by the European Research Council. His research focuses on the interplay of science, diplomacy, intelligence and surveillance operations, especially during the Cold War. His publications include *The Pontecorvo Affair: A Cold War Defection and Nuclear Physics* (University of Chicago Press, 2012) and *The Surveillance Imperative* (Palgrave, 2014, co-edited with Peder Roberts).

Jessica Wang
(PhD MIT, 1995) is Associate Professor of History at the University of British Columbia. Her publications include *American Science in an Age of Anxiety* (University of North Carolina Press, 1999), along with a range of articles and essays about the history of science, social science, and US state power.

Introduction

Cold War Science and the Transatlantic Circulation of Knowledge

Historical studies of science during the Cold War have received strong interest as the period recedes further into memory. A new term to designate its science, 'Cold War Science', was coined, and recent advances in historiography show a remarkable change in perspective.[1] The main narrative is no longer that of a Faustian tango between large government spending and Big Science projects, primarily focused on the USA or otherwise framed in a national context. It has become increasingly clear that, as a geopolitical state of affairs, the Cold War determined the circulation of knowledge on a global scale: the historiographical perspective is shifting towards transnational developments and how these shaped local decisions.

The tension between the global and the local is the subject of this book. The studies presented here mostly consider the US in interaction with its European partners. 'Cold War Science' is understood to refer to the practices of science that were specifically conditioned by Cold War circumstances: in contexts where the global strongly shaped the local, 'Cold War Science', rather than simply 'science in the Cold War', seems a particularly appropriate term. Yet, at the same time, the question whether the two terms and concepts may be distinguished at all in a meaningful way underlies many of the discussions presented in this book.

In the past decade, the thesis that has framed our understanding of the development of western European science vis-à-vis the US held that European science was reshaped according to American models, in the course of US efforts to reconstruct the Continent. The US often played the role of a benevolent hegemon creating opportunities to safeguard its influence, security and larger interests, such as the perception of success of capitalism as a social model.[2] While this thesis has been debated extensively,[3] a full understanding of its exact workings, and a consensus on how this view would apply across the board for the many different relationships involved, is yet to be achieved. Existing

1 See e.g. H. Heyck and D. Kaiser, "Introduction: New Perspectives on Science and the Cold War," *Isis* 101 (2010), 362–366; N. Oreskes and J. Krige (eds.), *Science and Technology in the Global Cold War* (Cambridge MA, 2014).

2 J. Krige, *American Hegemony and the Postwar Reconstruction of Science in Europe* (Cambridge MA, 2006).

3 See for example the articles in P.G. Abir-Am (ed.), *Post WW2 Transatlantic Science Policies*, special issue of *Centaurus* 52 (2010), 273–362.

 | DOI 10.1163/9789004264229_002

accounts have been mostly concerned with larger nations (e.g. Germany, France, and the UK), although recently the perspectives of small nations (like the Benelux and Scandinavian countries) have also received attention.[4] Some questions, however, remain: how did specific conditions of the Cold War, such as secrecy and a strong military interest in science, affect both the transatlantic and the intra-European circulation of knowledge? And given that European allies did not just react to a dominant hegemon, but also played an active role in their relationship with the Americans, how is their agency reflected in their science histories?

These open problems and imbalances in an otherwise increasing understanding of science in the Cold War period motivated a number of Dutch historians of science (Dirk van Delft, Jeroen van Dongen, Frans van Lunteren and Ad Maas, together with David Kaiser of MIT) to organize a workshop for more than fifty participants at the Lorentz Center at Leiden University, which took place 16th through 20th December 2013. The papers in this volume represent some of the best outcomes of this workshop.

What can we learn from these articles and the discussions in Leiden? Central in all, and certainly in the first article by Ron Doel, is the realization that to understand science during the Cold War, understanding the relation between science and the state is crucial. The state determines not only local conditions, but also the international contexts of science. It conditions when and how circulation of knowledge can take place. Doel recounts how the scientific ideal of internationalism was at odds with the desire of the United States to gather scientific intelligence, as well as with other domestic interests. However, it was possible for internationalism to regain currency when internationalist ideals were recast as serving the national interest; this was the case, for example, when their enforcement was expected to further the cause of democracy abroad. Doel's account also points to an immediate problem for historiography: many of the discussions took place under the veil of 'secrecy'. Can historians gain a sufficient grasp on how secrecy influenced the circulation of knowledge to write the history of Cold War Science? Even if it is possible to attain a sense for the role of secrecy at the policy making level, how can we understand its influence on the daily work of a scholar? These questions, and some initial answers, feature in several papers at the heart of this volume.

Simone Turchetti's article addresses questions about secrecy, science and historiography directly. He describes how the NATO Science Committee sought ways to circulate classified science, mostly of US origins, among its members

---

4 M. Heymann and J. Martin-Nielsen (eds.), *Perspectives on Cold War Science in Small European States*, special issue of *Centaurus* 55 (2013), 221–359.

without hampering security interests. Eventually, this process contributed to the identification of areas in science that required sponsorship from NATO. It also helped to declassify some knowledge, which in turn enabled new collaborations in the NATO context. Secrecy effectively gave the US a hand in what research was undertaken by its European partners: where the former lifted secrecy, the latter saw new research opportunities. Secrecy and its assertion in the work of the individual scientist is investigated in the article by Abel Streefland, which focuses on Dutch physicist Jakob Kistemaker and his ultracentrifuge program. Although initially working without any substantial security regulations, Kistemaker was forced to find new ways of continuing his project when confronted with American demands to classify it in 1960, purportedly due to fears of proliferation. In fact, the US also used secrecy stipulations to secure an altogether different kind of interest: it intended to slow European centrifuge research down, if not thwart the various programs altogether, to have its market share for enriched uranium remain unchallenged.

This course of action by the Americans mirrors that of a few years later, when the US wished to gain access to details of British centrifuge designs. The Americans wanted to ensure that the sharing of British designs with potential European partners did not violate secrecy arrangements between themselves and the British; unlike Dutch centrifuges, British centrifuges had been developed using US technology. UK scientists, however, saw this as an attempt at industrial espionage; indeed, American wishes for information were foremost motivated by the desire to retain their global leadership position in uranium production. This episode, described in great detail by John Krige elsewhere,[5] demonstrates, according to Krige, that knowledge in the Cold War circulated in a network with the US as a dominant pole in its center. It received feedback through its interactions with other nations: new knowledge was co-produced by exchanges in this asymmetric network—exchanges that depended on reciprocity. Studies in this book show that such reciprocity was actively sought by junior partners, and awarded and regulated by the United States. The possibility of such *quid pro quo*, to cite the title of the article by van Dongen and Hoeneveld, would lead to European investment in science at home to obtain knowledge from abroad. The Dutch in particular saw a need to invest in their national science in order not to be sidelined in knowledge exchanges. This was considered a pressing matter since the Netherlands was too small a country to expect any substantial progress if it worked in isolation, particularly in fields that required an unattainable scale and level of investment. In the end, the

5 J. Krige, "Hybrid Knowledge: The Transnational Co-Production of the Gas Centrifuge for Uranium Enrichment in the 1960s," *British Journal for the History of Science* 45 (2012), 337–357.

Cold War presented Dutch scientists with both challenges and a variety of opportunities, such as, for example, the Fulbright program (see the contribution by Giles Scott-Smith) and the creation of a major NATO research center in The Hague ('SHAPE Technical Center', see van Dongen and Hoeneveld).

Krige's contribution to the present volume once again draws our attention to the central role of knowledge circulation, and to the importance of determining its nature and motivations. In addressing these questions, particular attention needs to be paid to transnational perspectives: British and French nuclear programs were helped along by the US, but at a cost, as both countries were expected to align their foreign policy priorities more closely to those of the US in consequence. Such workings of exchange mechanisms can be pointed out at the global level of policy makers and politicians, but also at the local level, especially in the daily work of individual scientists that require international interaction to conduct their research. At the same time, power relations strongly prescribed the co-production of knowledge. Kistemaker's example is a case in point: the US, by imposing the classification of centrifuge knowledge, essentially sought to shut down its circulation among prospective European partners, and thus to render its development nearly impossible. Power relations could further be brought to bear either through direct financial incentives, and by offering access, or by withholding it through secrecy.

The Kistemaker case also exhibits another aspect of the circulation of knowledge: its conditioning not only by power relations, but also by cultural differences. The Dutch were truly taken by surprise by the American security requests. Their own lack of a formalized classification system reflected their trust in personal relations and the individual moral judgment of character. In the US, however, a rather more technocratic understanding of how to assess human relations and behavior was prevalent, as illustrated in Paul Erickson's account of the study of 'values'. The ways in which the US dealt with security issues and questions of proliferation reflect this, for example through the confidence placed in game theoretic methods.[6] Of course, the modest scale of Dutch science rendered its lack of formal security measures more manageable; however, sensitive centrifuge technology eventually reached Pakistan due to successful espionage activity of Abdul Qadeer Khan in Holland in the 1970s.

---

6 P. Erickson, J.L. Klein, L. Daston, R. Lemov, T. Sturm and M.D. Gordin, *How Reason Almost Lost Its Mind: The Strange Career of Cold War Rationality* (Chicago IL, 2013).

Dutch policy makers, as always, were acutely aware of the relatively small size of their country. In the Netherlands, as mentioned before, heavy investments in science were made because partnerships needed to be found.[7] The Dutch understood particularly well that, without knowledge of their own, "the US will think of Holland as of Portugal or Romania," according to Dutch-American astronomer Gerard Kuiper in a letter to prime minister Willem Schermerhorn (see the contribution of David Baneke). The articles in this volume show that the Dutch invested in centrifuge knowledge (Streefland), chemical warfare research (Roozenbeek), created the 'Rijksverdedigingsorganisatie' (or 'RVO'), a substantial institution for defense-related research (van Dongen and Hoeneveld), and strongly supported the Fulbright program (Scott-Smith). Indeed, Dutch researchers considered an inclusion of their efforts in international collaborations as one of the strongest markers of success. This applies to the ways in which G.J. Sizoo gauged RVO's role in the co-creation of NATO's 'SHAPE Technical Center', and chemist J. van Ormondt's valuation of his collaboration with the French, and the access he was given to British and American data; connections with the Americans were particularly highly valued. Invariably, collaborations only ensued if the Dutch could offer knowledge of their own. In the case of the Fulbright program, it was the adaptation of Dutch scientific and social elites to American values and practices that was tacitly traded. Thus, both knowledge and hegemony were co-constructed through circulation. Even though the US set the conditions of the Western alliance, peripheral actors were not without choices or opportunities: for them, producing knowledge at home could break open the secrecy regimes of others, and thus grant access to new chances. At the same time, knowledge as a contribution to the Western alliance allowed smaller countries to attain a more prominent spot within it.

The outlined mechanisms are also exhibited in the case of Danish researchers working in Greenland. Science was an important diplomatic instrument for the Danish state in navigating its complicated relationship with the US, which laid great claims on Greenland due to its strategic position in the northern Atlantic. Matthias Heymann, Henry Nielsen, Kristian Hvidtfelt Nielsen and Henrik Knudsen show that, through expanding their efforts in science in Greenland, the Danes tried to secure their hold on the island. The Americans, however, soon conducted research and defense activities in Greenland at such

7 F. Hoeneveld and J. van Dongen, "Out of a Clear Blue Sky? FOM, The Bomb, and The Boost in Dutch Physics Funding after World War II," *Centaurus* 55 (2013), 264–293.

a grand scale that the Danes effectively lost their overview of US activities, and their control of the situation. In the end, political leaders were forced to suspend particular scientific collaborations after an American airplane crashed on the island before the eyes of a highly critical Danish public. This was much to the regret of Danish scientists who feared for their relations with the US, as they anticipated an American interpretation of this move as a lack of reciprocity from the Danish side—an awkward response given the great liberties that the Americans had claimed for themselves in Greenland, in spite of strong Danish reservations; US hegemony in Greenland does not seem to have been a voluntary co-construction. If compared with the Danes, however, the Dutch would have envied the access and opportunities available to the Danes.

For co-production of knowledge to take place, not only people needed to circulate: data and ideas had to be shared, too. Weather modeling is one field that would not have succeeded without transnationally shared data, as Kristine Harper's article elucidates. Harper also points out an additional, vital aspect of 'Cold War Science': its hubris. Indeed, concrete programs for 'weather control' or the quantification of morality (Erickson), or the construction of nuclear powered cities under the ice cap in Greenland (Heymann et al.), show remarkable, excessive expectations for the promise of science that seem particular to the period. Such efforts are not just outgrowths of scientific fields that benefitted from Cold War circumstances, but they are expressions of a particular frame of mind.

David Baneke and Jessica Wang remind us that the immediate geopolitical East-West conflict not only, or not even necessarily, shaped science policy during the Cold War years. In the case of European countries, local economic development and intra-European economic and political competition were equally defining basic contexts. Nonetheless, relations with the US always loomed large in the minds of European policy makers, while surprisingly, it was in fact the Soviet Union that was mostly absent from their considerations, as Baneke shows. Wang argues that, on the global scale, science, and particularly social science, was often implicitly enlisted to retain power relations between the global center and the 'underdeveloped' periphery. At the same time, these relations were utilized in producing scientific knowledge. Furthermore, they remained firmly in place while the colonial powers were gradually replaced by the US-centered imperialist and developmental networks of the Cold War era. Frequently, the same ideas if not the same persons circulated, facilitated by the same institutions, as Wang points out. Again, as throughout this book, it becomes clear that an understanding of the transnational circulation of knowledge is vital to understand the workings of science in the Cold War period.

We thank the Lorentz Center and its staff, particularly Corrie Kuster and Henriette Jensenius, for making possible such a memorable and exceptionally well-facilitated workshop. We are also grateful to the workshop participants for the lively and most rewarding debate, and for the excellent atmosphere. We further thank the Lorentz Center sponsors and the sponsors who particularly co-financed our workshop for their support: The Royal Netherlands Academy of Arts and Sciences, Museum Boerhaave, the University of Leiden, the Netherlands Institute for Advanced Studies (NIAS), the Descartes Center for History and Philosophy of Science of Utrecht University and finally Stichting Physica. This book is partly financed by the Foundation for Fundamental Research on Matter (FOM), which is financially supported by the Netherlands Organization for Scientific Research (NWO). Finally, we thank Anke Timmermann for assistance with copy editing.

Jeroen van Dongen, Friso Hoeneveld and Abel Streefland
Amsterdam, 2015

PART 1

Secrecy and Science

CHAPTER 1

Scientists, Secrecy, and Scientific Intelligence: The Challenges of International Science in Cold War America

*Ronald E. Doel**

American science in the early Cold War embraced contradictory practices. This was particularly the case when scientific internationalism clashed with new demands for scientific intelligence. During and immediately after World War II, leaders of the US scientific community worked hard to institutionalize an unprecedented working relationship between science and government, aware that science had made unique contributions to the nation's military strength—a relationship they understood would also promote and expand resources available to their professions. Like most Americans, many US scientists had come to believe that American power was the world's most potent antidote to totalitarianism. They also believed that totalitarianism was fundamentally inconsistent with intellectual freedom and scientific integrity, whether the virulently racist form rooted in Nazi Germany or varieties of Stalinist-styled communism. For this reason, few American scientists had qualms about the state-science partnership that they were so instrumental in establishing, and few doubted that science was a promising vehicle to promote democratization within the Soviet Union, well equipped to maintain vital communications across the East-West divide.

Yet at the same time, leaders of the US military and the federal government, who were similarly aware of this remarkable new alliance between scientists and the nation, saw the production and circulation of scientific knowledge (and the circulation of scientists through travel) increasingly in national and strategic terms. Since scientific knowledge was now crucial to the development of new weapons systems and national security, were there any advantages to scientific internationalism and openness, except when these practices served

* I gratefully acknowledge support from the European Science Foundation's EUROCORES Programme "BOREAS: Histories from the North: Environments, Movements, Narratives," as well as National Science Foundation grants 0922651, 9112304, and SBR-9511867, the "Exploring Greenland: Science and Technology in Cold War Settings" project, Aarhus University, and the Tanner Humanities Center, University of Utah.

 | DOI 10.1163/9789004264229_002

the national interest? Like US scientists, many political leaders recognized ideological intrusions into the practice of science in totalitarian regimes. Trofim Lysenko's successful, Kremlin-backed campaign to crush Western genetics in favor of acquired Lamarckian characteristics was the best-known example. But they took away a different lesson than most scientists did: since there was little to gain from these foreign research traditions, the US had no reason to maintain the practice of scientific internationalism. Viewed through this lens, science was not superior to economics, politics, or culture. Rather, scientists needed to adhere, at all times, to policies and restrictions promoted by the state to serve its foreign policy interests.[1]

These two perceptions of scientific internationalism increasingly came into conflict early in the Cold War, particularly as US officials recognized that scientific intelligence had potential value that went far beyond the immediate needs of the military. Many American scientists grew concerned about the mutual relationship they had forged with the state, aware of the contradiction between tight control of information for national security and the scientific ideal of universalism, which demanded that scientists resist at all cost the interference of non-scientific interests (religions, states, or political or military alliances) in shaping scientific research or controlling communications between scientists.[2]

Tensions along these fault lines deepened from the late 1940s through the 1950s. But conflicts were especially evident in the creation (and practice) of new organizations, and particularly within the newly established Central Intelligence Agency and the US Department of State, tasked with handling scientific intelligence. While the CIA was the first new American agency in the

1 Allan A. Needell, *Science, Cold War and the American State: Lloyd V. Berkner and the Balance of Professional Ideals* (London, 2000); Ronald E. Doel and Zuoyue Wang, "Science and Technology," in Alexander DeConde, Richard Dean Burns and Frederik Logevall (eds.), *Encyclopedia of American Foreign Policy* (New York, 2001), 443–459; see also Peter Galison, "Secrecy in Three Acts," *Social Research: An International Quarterly* 77 (2010), 941–74. I thank Allan Needell in particular for many extended discussions on this issue; this article draws in part on prior joint research and writing. I also thank Kristine C. Harper for helpful criticism.

2 Ronald E. Doel and Allan A. Needell, "Science, Scientists, and the CIA: Balancing International Ideals, National Needs, and Professional Opportunities," *Intelligence and National Security* 12 (1997), 59–81, on p. 59; Michael Aaron Dennis, "Secrecy and Science Revisited: From Politics to Historical Practice and Back," in Ronald E. Doel and Thomas Söderqvist (eds.), *The Historiography of Contemporary Science, Technology, and Medicine: Writing Recent Science* (New York, 2006), 172–184; Paul Forman, "Into Quantum Electronics: The Maser as 'Gadget' of Cold War America," in Paul Forman and José M. Sánchez-Ron (eds.), *National Military Establishments and the Advancement of Science and Technology* (Dordrecht, 1996), 261–326.

early Cold War to mount sustained efforts to obtain and evaluate scientific intelligence, actions by the Department of State caused American scientists far greater concern by the late 1950s, as State Department restrictions on scientific internationalism threatened the autonomy, professional prestige, and international influence of US scientists. But the US scientific community, a broad spectrum of disciplinary communities from mathematics to geology, from molecular biology to ecology, was not monolithic.[3] Physics was the most international of natural science disciplines in outlook and practice, even before its role in creating the atomic bomb thrust physicists onto the world stage, and physicists worried deeply about secrecy and restrictions on scientific internationalism. However, it is also now clear that a significant number of American scientists, particularly chemists, saw no problem with aligning science with state interests, complicating prior narratives of this conflict.[4]

This article reviews briefly how American scientists sought to resolve fundamental contradictions between national demands and international traditions; and how, with varying degrees of success, American scientists became involved in the formulation and execution of their government's policy. More than a run-of-the-mill domestic policy struggle, state attempts to regulate the international relations of science illuminate central conflicts over autonomy and loyalty, as state and private spheres of authority were negotiated and defined. State influences on the practice of international science in the Cold War were wide-ranging. While the best-known examples involve the restriction of physics data useful for constructing nuclear weapons, they also included the classification of many kinds of geophysical data important for weapons development: for instance, measured ocean depths (of strategic value in anti-submarine warfare), and Soviet environmental sciences research in the Arctic (where Pentagon leaders thought World War III might be waged, given the Arctic's strategic location).[5] Also relevant were secret Department of State instructions to US scientists to use their

---

3 Ronald E. Doel, Dieter Hoffmann and Nikolai Krementsov, "National States and International Science: A Comparative History of International Science Congresses in Hitler's Germany, Stalin's Russia, and Cold War United States," *Osiris* 20 (2005), 49–76.

4 Daniel J. Kevles, *The Physicists: The History of a Scientific Community in Modern America* (New York, 1977); Jessica Wang, *American Science in an Age of Anxiety: Scientists, Anticommunism, and the Cold War* (Chapel Hill, NC, 1999).

5 Ronald E. Doel, Tanya J. Levin and Mason K. Marker, "Extending Modern Cartography to the Ocean Depths: Military Patronage, Cold War Priorities, and the Heezen-Tharp Mapping Project, 1952–1959," *Journal of Historical Geography* 32 (2006), 605–26; Shelagh D. Grant, *Polar Imperative: A History of Arctic Sovereignty in North America* (Vancouver, 2010); Ronald E. Doel, Robert Marc Friedman, Julia Lajus, Sverker Sörlin and Urban Wråkberg, "Strategic Arctic

voting authority within government-sponsored international scientific bodies in order to restrict the participation of scientists from the 'unrecognized' nations of East Germany and mainland China—demands that deeply troubled many (but not all) US scientists.[6] Finally, they also included the placement of leaders of covert scientific activities in 'cover' positions: for instance, the second-in-command at the US National Bureau of Standards in 1947, the chemist Wallace Brode, was actually organizing scientific intelligence within the CIA. At the same time, State Department staff placed science attachés within US embassies in Western Europe, their status as observers and diplomats clouded by perceptions (sometimes accurate) that various Washington agency leaders wanted them to serve as intelligence agents.[7] We now know that the institutional landscape of Cold War science was a kind of fun-house mirror image: partially an accurate reflection, yet often distorted, concealing important, subterranean networks of connections. These issues affected all fields of the natural sciences, but our historical assessment of this rich and revealing landscape is far from complete.

Building a 'Science in Black' in Cold War America: Assessing Loyalties and Foreign Threats

American scientists did not fail to miss growing state interest in international scientific activities before the outbreak of World War II. Indeed, in the 1930s, US researchers were well aware that the actions of foreign states had begun to limit international scientific congresses, the primary point of intersection between the national and international dimension of scientists' activities. In 1937 the Soviet Union had rescinded an invitation for the Seventh International Genetics Congress in Moscow, sensing potential propaganda benefits would fall short, and plans to host an international aeronautics con-

---

Science: National Interests in Building Arctic Knowledge—Interwar Era through the Cold War," *Journal of Historical Geography* 44 (2014), 60–80.

6 Doel, Hoffmann, and Krementsov, "National States and International Science."

7 Ronald E. Doel, "Scientists as Policymakers, Advisors, and Intelligence Agents: Linking Contemporary Diplomatic History with the History of Contemporary Science," in Thomas Söderqvist (ed.), *The Historiography of Contemporary Science and Technology* (Amsterdam, 1997), 215–44; Ronald E. Doel, "Does Scientific Intelligence Matter?" *Centaurus* 52 (2010), 311–22; Needell, *Science, Cold War, and the American State*; see also Simone Turchetti, Néstor Herran, and Soroya Boudia, "Introduction: Have We Ever Been Transnational? Towards a History of Science Across and Beyond Borders," *British Journal of the History of Science* 45 (2012), 319–36.

gress in Germany faltered when Nazi Germany prohibited attendance by Jews, both domestic and foreign.[8] American scientists simply felt this would not become an issue in the United States. Participating in international scientific meetings boosted America's standing in the world and "carried the march of civilization forward," Wilbur J. Carr, then Assistant Secretary of State, had declared in 1937. They felt confident the classic ideological foundation of scientific internationalism would hold; that is, "the participation of the nation in the scientist's fame spares the scientist any conflict between advancing his science and advancing the interest of his nation."[9]

Careful observers nevertheless recognized that changes were occurring by the end of the 1930s. A key player in these changes was the Federal Bureau of Investigation (FBI), first established in the early twentieth century but only assuming its modern role as an independent agency within the Department of Justice in 1935. By 1940, as US leaders sensed that American involvement in World War II was increasingly likely, FBI director J. Edgar Hoover ramped up efforts to determine the loyalty of American scientists, aware that scientists had to be vetted by military security boards who had been tasked with clearing suitable candidates for wartime leadership roles involving science. Hoover, however, regarded the political loyalties of scientists as suspect, and their inclination to support internationalism as inherently subversive. To assess the loyalty of scientists, Hoover and his agents employed a number of 'litmus tests' to identify seditious tendencies, including participation in organizations the FBI deemed sympathetic to Communism. Hoover's approach soon had real consequences. When Vannevar Bush, the MIT engineer and head of the National Defense Research Committee, sought to appoint the eminent physical chemist Roger Adams of the University of Illinois to take charge of wartime US chemistry programs in July 1940, Bush discovered that Adams had been blacklisted in part for supporting the "Lincoln's Birthday Committee for the Advancement of Science"—for Hoover a Communist-front organization, since it had supported Franco's opponents in Spain. As it turned out, the Lincoln's Birthday Committee for Democracy and Intellectual Freedom (the group to which Adams had lent his name) had gained support from thousands of academics, conservative as well as liberal, primarily for opposing distorted and prejudicial definitions of 'race' in American textbooks, a major concern given the increasingly ugly racial policies in Nazi Germany. Hoover ultimately withdrew the FBI's objections; Adams was appointed; and, during the war, Adams

8 Nikolai Krementsov, *International Science Between the World Wars: The Case of Genetics* (New York, 2005); Doel, Hoffmann, and Krementsov, "National States and International Science."
9 Doel, Hoffmann, and Krementsov, "National States and International Science," p. 51.

helped lead a successful effort to develop synthetic rubber materials to replace those that were no longer available from Indonesia after Japanese forces conquered these islands. Had Bush not succeeded, Adams would have been kept out of crucial developments in wartime America, impacting American chemistry as well as national security. It is not an exaggeration to say that Hoover's anti-cosmopolitan worldview helped institute secrecy practices in American science even before World War II came to America with the Japanese bombing of Pearl Harbor in Hawaii in December 1941.[10]

World War II did not end scientific internationalism, but rather recast it into Allied and Axis spheres, with neutral nations serving as exchange points. Yet what was particularly new in the early 1940s was the extent to which states became concerned about progress in virtually all fields of research, funded substantial, large-scale research and development programs, and sought to classify new knowledge gained through scientific intelligence. The prime example of this was the Manhattan Project, which led to the development of the atomic bomb, but it was no less the case for all wartime scientific projects, including the development of radar, the exploration of effective ways to mass-produce penicillin, and the discovery of the deep sound channel in the ocean (which aided antisubmarine warfare, making it a crucial military advance). Secrecy became deeply integrated into knowledge production in each of these undertakings. For instance, the Bureau of Economic Warfare, charged with assessing the geographic distribution and quantities of strategic minerals throughout the world, recruited scientists with backgrounds in geology and geophysics, some of whom were previously employed by companies that had long made such estimates for proprietary reasons.[11] When scientists joined the US military in the Alsos mission in late 1944, sweeping behind Allied forces as they moved eastward across France and Germany towards Berlin with the mandate to interview captured German scientists about their wartime research achievements, they were well aware that the information they produced would be held tightly by the state.[12]

---

10 Ronald E. Doel, "Roger Adams: Linking University Science with Policy on the World Stage," in Lillian Hoddeson (ed.), *No Boundaries: University of Illinois Vignettes* (Champaign-Urbana, IL, 2004), 124–44; Wang, *American Science in an Age of Anxiety;* Fred Jerome, *The Einstein File: J. Edgar Hoover's Secret War against the World's Most Famous Scientist* (New York, 2003).

11 Dennis, "Secrecy and Science Revisited;" Graham D. Taylor, "The Axis Replacement Program: Economic Warfare and the Chemical Industry in Latin America, 1942–1944," *Diplomatic History* 8 (1984), 145–64.

12 A helpful recent review is John D. Hart, "The ALSOS Mission, 1943–1945: A Secret U.S. Scientific Intelligence Unit," *International Journal of Intelligence and Counterintelligence* 18 (2005), 508–37.

But it was in the immediate post-war period that the US state intensified efforts to gain knowledge about scientific activities beyond its borders. This took several forms. One, covert in design and practice, involved creating new professional channels and agencies to collect scientific intelligence.[13] A second, and subsequent, effort focused on finding ways to integrate international science activities with foreign policy aims in the US Department of State. These initiatives took place alongside continued efforts by a much-expanded FBI, still under the directorship of J. Edgar Hoover, to gather copious details on the state of foreign scientific laboratories and individual scientists (primarily obtained through interviews by US scientists who traveled overseas), along with much-expanded efforts to vet the loyalties of a significant group of US scientists. Together these activities revealed unprecedented state interest in international scientific activities, the result of which was a dense web of largely secret information, available only to individuals at top echelons of the state, which simply had not existed fifteen years previously when the presidential administration of Franklin D. Roosevelt began.

Creating a scientific intelligence branch within what became the Central Intelligence Agency seemed a nearly insurmountable challenge to many US academics, given the models of scientific practice then commonly accepted. In 1943 the famed sociologist of science Robert Merton had begun to articulate what he called the fundamental norms of science, later denoted by the acronym CUDOS: communalism, universalism, disinterested, and organized skepticism. After World War II, Merton's formulation was embraced by American political leaders as a promising means to advance US foreign policy. James B. Conant, the president of Harvard University, wartime science administrator, and one of the leading educators of his generation, declared in 1945 that free inquiry would aid capitalism, industrial development, and democratization: that is, science could flourish only in democratic states, where (in contrast to totalitarian regimes) free inquiry was encouraged and supported. These widely held views seemed incompatible with state hopes to create robust structures to gather scientific intelligence. Indeed, through the onset of World War II, scientists had resisted attempts to curtail international scientific exchanges except in times of actual warfare.[14]

---

13 Doel and Needell, "Science, Scientists, and the CIA;" see also Jeffrey Richelson, *Spying on the Bomb: American Nuclear Intelligence, from Nazi Germany to Iran and North Korea* (New York, 2007).

14 Michael Aaron Dennis, "Historiography of Science: An American Perspective," in John Krige and Dominique Pestre (eds.), *Science in the Twentieth Century* (New York, 1997), 1–26; James G. Hershberg, *James B. Conant: Harvard to Hiroshima and the Making of the Nuclear Age* (New York, 1993); see also Ronald E. Doel, "Science, Religion, and the Modern State:

Yet leaders of the Pentagon's Research and Development Board (RDB), a postwar institutional innovation that became the principal intersection between leading American scientists and their military counterparts, did just this, expanding wartime military scientific intelligence efforts in new directions.[15] In January 1947 Vannevar Bush, by then director of the RDB, began working closely with General Hoyt Vandenberg, head of the Central Intelligence Group (the predecessor organization of the CIA) on the problem of gathering scientific intelligence sufficient to meet the needs of the RDB, since that body was charged with determining what scientific and technical advances were needed to maintain US national security. Bush and fellow-RDB leaders chose Wallace R. Brode to lead this effort; Brode, a University of Illinois PhD, had taught at Ohio State University before becoming director of the Science Department of the Naval Ordnance Test Station at Inyokern, California, where he had developed plans for scientific intelligence gathering. Since any new professional appointment would require him to resign from his tenured position in the Department of Chemistry at Ohio State (having exhausted his available leaves there), Brode was officially appointed second-in-command at the National Bureau of Standards—a cover position for his covert activities.

Brode faced extraordinary challenges in creating a functioning branch of scientific intelligence within the nation's emerging powerhouse for intelligence gathering and covert operations. Indeed, he stayed in this position for just one year, and his efforts were subsequently judged a failure. One of Vannevar Bush's lieutenants bluntly declared in late 1948 that the "RDB has received substantially no scientific intelligence of the type wanted from CIA."[16] Yet Brode's tenure at the CIA, and particular challenges he faced in creating what ultimately became a stable system of national scientific intelligence gathering, revealed a daunting number of institutional and professional difficulties that stood in the way of any American scientist so tasked. One was the bewildering array of military and governmental agencies already collecting scientific information, including the Army, Navy, the newly formed Atomic Energy Commission, and a myriad of other cabinet departments. Few of them were willing to share

---

A View from the History of Science," in Volker Depkat and Jürgen Martschukat (eds.), *Religion and Politics in Europe and the United States: Transnational Historical Approaches* (Washington, DC, 2013), 163–86, on pp. 165–8.

15 Then known as the JRDB, the 'Joint' RDB (for Army and Navy) disappeared after the 1947 reorganization that transformed the US Department of War into the Department of Defense, with the Air Force added as a new branch.

16 Doel and Needell, "Science, Scientists, and the CIA," quoted on p. 63.

information with the CIA. Another problem was a lack of support within the CIA, including from its first director, Admiral Roscoe Hillenkoetter, who had little appreciation of the particular challenges of gathering this information, and had left Brode floundering when other CIA branch heads challenged his authority. Finally, Brode saw dangers in the pattern of scientific intelligence gathering that he had inherited, which emphasized six categories of 'immediate interest': aeronautics, atomic energy, biological warfare, chemical warfare, communications and electronics, and guided missiles. While these categories had institutional analogs within the RDB, Brode believed that scientific intelligence worked best if gathered within traditional disciplinary frameworks (physics, chemistry, biology, and the like) so that potential threats not marked as 'targets' could be identified. Recreating the university model within the CIA—a proposal he successfully pushed through—was Brode's most enduring contribution to the office, and one that helped recruit trained scientists to the covert realm of gathering and interpreting scientific intelligence.[17]

Scientific intelligence did not become a robust branch of the CIA until September 1949, just after the Soviet Union exploded its first atomic weapon. Scientific intelligence experts within the Agency had predicted the first possible date for the Soviet bomb at the end of 1950, an embarrassing sixteen months later than the actual first Soviet blast, an intelligence failure that unsettled Washington. Then CIA scientific intelligence chief Willard Machle, an MD by training who had directed the Armored Force Medical Research Laboratory in Kentucky, quickly drafted a four-page memo on the inability of the Office of Scientific Intelligence (OSI) "to Accomplish its Mission," circulating it even as radioactive debris clouds from the Soviet test drifted through the atmosphere.[18] Machle's sharp criticism of the refusal of military intelligence agencies to share scientific intelligence worked: Hillenkoetter boldly backed Machle, and the relationship between the Research and Development Board and the CIA warmed. Machle also succeeded in recruiting additional scientists to the OSI, an effort aided by the Communist take-over of Czechoslovakia in 1948, by heightened tensions between the US and the Soviet Union, and by increased concern among American scientists about ideological intrusions into Soviet science; this last aspect is exemplified by Trofim Lysenko's campaign against Soviet geneticists and published attacks on Western scientists as "learned lackeys of capitalism." By the time the Korean War broke out in late 1950, many US scientists had come to regard the Soviet Union as a credible

17 Ibid.
18 Ibid., p. 67.

threat to American national security, believed that Communist Party actions had distorted the Soviet scientific community, and accepted that scientific intelligence work could aid the standing of US science while advancing democratic agendas.[19]

Even with these changes, however, the CIA's OSI—led by medical experts and chemists through the early 1950s, many employed in industry and hence already adept in keeping trade and proprietary secrets—did not succeed in providing the range of scientific intelligence its state clients most desired. Years later, in 1976, the famous US Senate-led Church Committee probe of the Agency concluded that CIA intelligence specialists had sought to do everything, and thus, from the perspective of policy-makers, contributed almost nothing. There was truth to this charge: OSI findings did not always serve the needs of White House officials, members of Congress, or leaders of government agencies. For instance, OSI analysts, in one of their first major reports, carefully reviewed the number of published abstracts in Soviet chemistry from 1941 forward. Their key finding: this number had plummeted over 50 percent, revealing how much new research was 'born classified.' But this intelligence finding did not answer the question of greatest concern to state policymakers: what types of novel chemical weapons might Soviet chemists produce? Other scientific intelligence findings, however, did prove more relevant to foreign policy concerns. In 1949, analysts produced OSI 1/49, "An Estimate of Swedish Capabilities in Science:" a finely detailed 64-page report that evaluated Swedish scientists, research institutions, universities, and technological programs, including highly classified wind tunnel and pulse jet engine research. Almost certainly the report was prompted by the creation of NATO: in contrast to its Scandinavian neighbors Norway and Denmark, which became charter members of NATO, Sweden chose to continue its policy of neutrality, making it potentially vulnerable to Soviet invasion. The report included a list of top Swedish scientists to deny to Soviet invaders—precisely the kind of information desired by the National Security Council and military agencies.[20] At the same time, OSI 1/49 revealed the continued fragmentation of scientific intelligence-gathering efforts within the United States early in the Cold War. Likely because scientific intelligence involving the earth sciences and geophysics remained securely within the provenance of Naval intelligence through the 1940s, CIA science analysts made no mention of the University of Stockholm's

---

19 Doel, Hoffmann, and Krementsov, "National States and International Science;" see also Nikolai Krementsov, *Stalinist Science* (Princeton, NJ, 1997); Ethan Pollock, *Stalin and the Soviet Science Wars* (Princeton, NJ, 2006).

20 Doel and Needell, "Science, Scientists, and the CIA."

Hans W. Ahlmann, one of the West's leading glaciologists and climatologists. In 1947, breaking from a US lecture tour, Ahlmann had visited a meeting of the RDB to warn US military leaders that polar melting was likely to offer the Soviet Union advantages (agricultural productivity would increase, and northern ports would remain ice-free for ever longer periods). CIA Arctic branch specialists would soon trumpet the importance of climate change and physical environmental sciences knowledge for US national security, as well as Ahlmann's centrality in this research. But for clients of the CIA's OSI, even the university model of assessing major potential threats did not work when entire university departments and disciplines were kept from OSI purview.[21]

While integrating scientific intelligence into the CIA proved difficult, it was harder still to engineer a role for science in the Department of State—a matter with greater consequences for American science, as it turned out, by the early 1960s.

Because the State Department had stagnated during World War II, Department leaders, including James F. Byrnes and George C. Marshall, the first two Secretaries of State to serve President Harry S. Truman, began planning how this long-influential unit could regain its traditional place at the center of planning and execution of American foreign policy.[22] The still fluid post-war relationship between science and government, these leaders further recognized, would also impact the Department. This was precisely what many leading US scientists hoped would happen, believing that State could become a strong voice in science policy, providing a counterweight to increased military influence on international scientific activities. Their hopes were buoyed in 1947 when a high-level report sought by the Truman Administration, and produced by one of the President's senior assistants, John R. Steelman, argued

21 Ahlmann was a skilled diplomat and politician as well as a scientist; he ultimately served as Sweden's ambassador to Norway. While wary of possible US efforts to dominate and restrict Western Arctic research, he was adroit in recognizing what issues might stimulate US military authorities to undertake significant scientific studies; on these issues see Doel, Friedman, Lajus, Sörlin, and Wråkberg, "Strategic Arctic Science," and Sverker Sörlin, "Ice Diplomacy and Climate Change: Hans Ahlmann and the Quest for a Nordic Region Beyond Borders," in Sverker Sörlin (ed.), *Science, Geopolitics, and Culture in the Polar Region: Norden Beyond Borders* (Farnham, 2013), 23–54.

22 Charles E. Neu, "The Rise of the National Security Bureaucracy," in Louis Galambos (ed.), *The New American State: Bureaucracies and Politics Since World War II* (Baltimore, MD, 1987), 85–108; Michael J. Hogan, *A Cross of Iron: Harry S. Truman and the Origins of the National Security State, 1945–1954* (New York, 1998); and Brian J. Balogh, "Reorganizing the Organizational Synthesis: Federal-Professional Relations in Modern America," *Studies in American Political Development* 5 (1991), 119–72.

that "the security and prosperity of a nation depend today as never before on the rapid extension of scientific knowledge. So important, in fact, has this extension been to our country that it may reasonably be said to be a major factor in national survival." A subsequent high-level committee, endorsing the Steelman Report, added that it was in the US national interest "to lend every possible aid to research and development in all those countries willing to enter wholeheartedly into cooperation with us."[23] Accepting the arguments for the importance of science to the nation's prosperity, Truman made two important decisions in early 1949. Publicly, in his State of the Union speech, Truman proclaimed what became known as the 'Point IV' initiative, making American experience and technical know-how available to the developing, newly decolonized nations of the world. Privately, that same month, Truman formally approved National Security Council Intelligence Directive 10, making the Department of State primarily responsible "for the collection abroad for all government agencies of information in the basic sciences," excluding only what defense agencies would secure through their own channels. Yet State, with its staff largely trained in the liberal arts, was ill-equipped to handle either task: grafting scientific intelligence onto State's already large domestic and overseas missions posed formidable challenges that initially defied solution.[24]

To address this problem, Department of State leaders soon began devising new strategies to integrate scientists into foreign policy planning. In part they looked outside the Department for help, turning to Lloyd V. Berkner, a senior geophysicist at the Carnegie Institution of Washington who had participated in numerous high-level studies of national security planning, leaving him adept at walking the thin line that separated the scientific community from national security and military demands. His review, "State Department Responsibilities in the Field of Science," later known as the Berkner Report, reiterated the advantages of promoting international science: doing so would benefit the world's economies and increase standards of living, while simultaneously advancing American national security interests. "[T]he international science policy of the United States must be directed to the furtherance of understanding and cooperation among the nations of the world," Berkner declared, "to the

23 Quoted in *President's Scientific Research Board*, vol. 1 (Washington, DC, 1947), p. 3 and in Needell, *Science, Cold War and the American State*, p. 136; see also Allan A. Needell, "From Military Research to Big Science: Lloyd Berkner and Science Statesmanship in the Postwar Era," in Peter Galison and Bruce Hevly (eds.), *Big Science: The Growth of Large-Scale Research* (Stanford, 1992), 290–311 and David M. Hart, *Forged Consensus: Science, Technology, and Economic Policy in the United States, 1921–1953* (Princeton, NJ, 1998).

24 Needell, *Science, Cold War, and the American State*, p. 138.

promotion of scientific progress and the benefits to be derived therefrom, and to the maintenance of that measure of security of the free peoples of the world required for the continuance of their intellectual, material, and political freedom." Among the solutions Berkner proposed was placing scientific attachés in various US embassies throughout Western Europe. Taking this step, Berkner asserted, would yield "certain definite benefits which are highly essential to the security and welfare of the United States, both generally and with respect to the progress of science."[25]

Berkner's widely distributed 1950 report reflected the views of many members of the US scientific community. There is little doubt that Berkner deeply believed what he wrote. Yet his published report was only part of the story. Very much aware of more the sensitive aspects of science and foreign relations raised by Truman's NSCID 10 directive, Berkner also prepared a Secret Supplement to his Department of State report. It is clear he considered this covert report quite significant: he advised its readers that the unclassified portion "should be considered a cover for the classified section." Here, Berkner strongly recommended that additional scientific personnel be placed within State to monitor international scientific developments. Doing so, he argued, would aid "the private international flow of scientific information" to support US aims, a crucial matter since scientific intelligence from Austria, Germany, and Great Britain was largely absent. In crafting the Secret Supplement, Berkner was walking a delicate line, serving as a middleman between the national security bureaucracy of the state and the professional community of scientists. Berkner saw no large contradiction in these roles; like many of his colleagues, he perceived no inconsistency between the goals of science, the goals of the international community of free people, and the interests of America.[26] It is perhaps not coincidental that, in the same month as the Berkner Report appeared, Berkner became one of the small circle of planners for what became the International Geophysical Year. The IGY, while ultimately becoming one of the largest international scientific undertaking of all time, was also an activity that US planners saw as serving vital American national security needs, from establishing the overflight principle for artificial satellites to acquiring vital physical environmental science data from around the globe necessary for weather forecasting and military activities.[27]

---

25 Ibid., p. 299.

26 Needell, "From Military Research to Big Science," pp. 300–2.

27 Walter A. McDougall, ... *The Heavens and the Earth: A Political History of the Space Age* (New York, 1985); Roger D. Launius, James Rodger Fleming, and David H. DeVorkin (eds.), *Globalizing Polar Science: Reconsidering the International Polar and Geophysical Years*

By 1952, the final year of the Truman Administration, State Department involvement in international science grew considerably. It placed science advisors in London, Stockholm, Bern, Bonn, and Paris, nearly a dozen scientists altogether, most of them academics on temporary leaves from their home universities. Several of these appointments proved challenging, in part because Western European scientists—even though unaware of Berkner's classified Secret Supplement—had grown suspicious of American intentions, seeing US science attachés as espionage agents. In June 1953, the communist publisher "Editions de la Nouvelle Critique" published a fiery 80-page polemic called *Un plan U.S.A. de mainmise sur la Science* [The American Annexation of Science], rebutting Berkner's *Science and the State Department* report with the charge that state-sponsored international activities were simply intended to steal Western European ideas to aid America. (Frederic Joliot-Curie, the prominent French nuclear physicist, had similarly denounced the attaché program as "l'activité d'espionage.")[28] Bonn's two science attachés, faculty members from Stanford University (William W. Greulich) and the University of Minnesota (Richard T. Arnold), noticed their telephones never rang; later, a German chemist informed Arnold that "most Germans regard you as a spy."[29] Greulich and Arnold had sought to erect a wall between themselves and the intelligence activities that were underway at the embassy; they were aided by then-US ambassador to West Germany, James B. Conant. Nevertheless, the incident revealed the extent to which evolving American definitions of scientific internationalism worked less well in Western Europe than science leaders such as Berkner had anticipated.[30] Indeed, the incident hinted at major challenges that were still to come.

(New York, 2010); Ronald E. Doel, "Constituting the Post-war Earth Sciences: The Military's Influence on the Environmental Sciences in the USA after 1945," *Social Studies of Science* 33 (2003), 635–66; Fae Korsmo, "The International Geophysical Year of 1957 to 1958," *Science, People, and Politics* 2 (2007), accessible at http://www.gavaghancommunications.com/sppkorsmoigy.html [accessed 31 May 2014].

28 Doel and Needell, "Science, Scientists, and the CIA," on p. 69.

29 Doel, "Does Scientific Intelligence Matter?" pp. 314–5.

30 On American practices involving science in early Cold War Europe see John Krige, *American Hegemony and the Postwar Reconstruction of Science in Europe* (Cambridge, MA, 2008) and John Krige, "Atoms for Peace, Scientific Internationalism, and Scientific Intelligence," *Osiris 21* (2006), 161–81.

Intensified Debates over Scientific Internationalism: Pre- and Post-Sputnik Challenges

In the 1950s, turmoil over the practice of scientific internationalism and scientific intelligence within the United States increased. One reason was that leading users of scientific intelligence within the federal government, including the Research and Development Board, found both the quality and the quantity of this information still lacking, despite the comprehensive systems that had been put in place in the CIA, military agencies, and the Department of State. Out of this frustration, the US government created another means of collecting foreign intelligence, including scientific intelligence: in 1952, Harry Truman instructed the Secretary of Defense, via the National Security Council, to create the National Security Agency (NSA). The NSA was charged with gathering information through signals intelligence, by careful monitoring of electronic communications around the globe. Like the science directorates in the CIA and the Department of State, the NSA would long outlive the Cold War that brought it into being.[31]

A more pressing issue at the time, however, was the still-unresolved debate about the proper role of international science as an element of US foreign policy. By the early Cold War, the concept that had prevailed within the American scientific community of the 1930s and Roosevelt's Department of State—that scientific internationalism was an activity that reflected glory to the nation—had long since vanished as the relation of science to military aims had become increasingly clear. Instead, the debate now focused on whether scientific internationalism benefitted US national security and thus needed to be encouraged (a position favored by CIA leaders), or whether international science was one of many cultural activities that needed to be subsumed under larger US foreign policy aims, to bring clarity to the existing East-West divide.

While, during the last term of Harry Truman's presidency, many American scientists were guardedly optimistic that the state would promote scientific internationalism as a benefit to US national security, by 1953, the year Dwight D. Eisenhower became the 34th President of the US, leaders of the US scientific community grew increasingly concerned. Although Eisenhower was a political moderate, his election placed conservatives in influential posts for the first time since Roosevelt's election in 1932. In the Department of State, headed by the hawkish, conservative John Foster Dulles, new appointees, deeply suspicious of US scientists who maintained contacts with scientists behind the Iron

31 James Bamford, *Body of Secrets: Anatomy of the Ultra-Secret National Security Agency* (New York, 2007).

Curtain, vigorously enforced the McCarran Act of 1950 (passed over Truman's veto), which allowed the state to bar entry to "politically suspect individuals." Their suspicions grew after Julius and Ethel Rosenberg were executed for sharing nuclear secrets with the Soviets that year—the same year in which unnamed insiders attacked the Department of State's fledgling science office in a note published by the sympathetic *U.S. News and World Report* labeling this office an "out and out stink hole of Communists." While the journal later printed a retraction, Dulles allowed the science office to wither, and the number of science attachés overseas dwindled to zero.[32]

Another concern felt by US scientists was that, while Eisenhower's Washington was making scientific internationalism more difficult in practice, international activity was picking up elsewhere in the world. After Stalin's death in 1953, Kremlin leaders—seeing an advantage in greater international engagement—began bold diplomatic initiatives. These included promoting cross-border scientific activities, as the Soviet Union joined several international scientific unions. One year later, as part of what became known as the 'Soviet offensive,' Soviet leaders hosted a lavish international rededication ceremony of the Pulkovo Observatory (destroyed in the siege of Leningrad in 1941), and invited world astronomers to plan an international meeting in their country with no political restrictions. These developments made American scientists anxious: if US scientists were barred from international scientific activities, Eastern Bloc scientists might gain control of the international scientific unions, reducing US influence while limiting the ability of Americans to remain at the cutting edge of their disciplines—particularly as Western European scientific institutions rebounded after World War II. Yet US visa denials multiplied. By 1955 four international scientific unions that had scheduled meetings in the US, aware of the American unwillingness to admit scientists with socialist or communist leanings, thought better of these plans, and relocated them to other nations. Leading American scientists worried that these limitations would damage what one called "incipient scientific relations between East and West."[33]

Yet Department of State officials, with tacit endorsement from the Eisenhower White House, continued to attack scientific internationalism through the months preceding the launch of Sputnik—convinced that privileging the concept of a global scientific community undermined the clarity

---

32 Quoted in Doel, Hoffmann, and Krementsov, "National States and International Science," pp. 67–8.

33 Quoted in Doel, Hoffmann, and Krementsov, "National States and International Science," on p. 68; see also Krige, *American Hegemony*.

of US foreign policy, including support for Formosa (Taiwan) over Mainland China, and West Germany over Soviet-dominated East Germany. In late 1956, through a set of covert communications kept from the view of American scientists, US diplomats secretly encouraged members of Taiwan's thinly populated, largely threadbare earth sciences community to demand the island nation represent China in the International Geophysical Year of 1957–1958 instead of Mainland China (where many more qualified scientists were active in related research projects). When Taiwanese scientists ultimately pursued their claim to represent China in international scientific programs, the Department of State, still under the forceful leadership of John Foster Dulles, instructed US scientists to support their colleagues in Taiwan. While Western researchers scrambled to find a compromise, deeply concerned about losing access to the knowledge and potential contributions of mainland Chinese scientists, Mao Tse-tung [Zedong] angrily withdrew his nation from the IGY after Taiwan (ruled by Chiang Kai-shek) maintained its demand, thereby denying Chinese scientists access to the largest international science undertaking of the twentieth century. It marked a low moment in Cold War scientific internationalism: communist China's pullout from the IGY did not result from intra-Chinese squabbles, as Western scientists then suspected, but from deliberate diplomatic interventions by the US Department of State—actions that put it at odds with the Office of Scientific Intelligence in the CIA, which favored more universal scientific activities (to better keep tabs on scientific research in communist China).[34]

Leading US scientists, including officers of the National Academy of Sciences (NAS), sought ways to reassert the advantages of scientific internationalism to the White House, to advance science as well as US foreign policy aims. By the mid-1950s, NAS president Detlev Bronk, increasingly dismayed by the failure of the US Department of State to integrate scientists into foreign policy, concluded that Dulles did not intend to implement the Berkner Report recommendations. Like Berkner, he was troubled by the rising number of foreign scientists denied entry to the US, and the limitation this placed on hosting international scientific congresses. Bronk nevertheless urged a cautious approach, arguing that "the long range interests of the nation will be better served by our sustained efforts to bring about a general change in the climate of opinion and action at the policy and operating levels of the Government than by strongly worded public statements of an Academy

34 All quoted in Doel, Hoffmann, and Krementsov, "National States and International Science," on p. 69.

position in individual cases."[35] The unexpected launch of Sputnik in October 1957 gave Bronk an opportunity to address what he termed "the evolving interactions of science with our national life in all its aspects," as criticism rose that Dulles might have anticipated Sputnik if State's science attachés had remained in place. Subsequently Dulles agreed to reinvigorate the science office, and authorized the securing of a new State Department Science Advisor. The individual soon appointed to this role was Wallace R. Brode, the physical chemist and Associate Director of the National Bureau of Standards who, one decade previously, had been tapped as the founding director of the scientific intelligence branch within CIA.[36]

Many US scientists who were anxious about the future of scientific internationalism applauded these developments. Bronk, Berkner, and their colleagues initially rejoiced in Brode's appointment, convinced that a solid consensus existed within the scientific community about the proper balance between openness and intelligence gathering in international science, and equally confident that Brode would advocate for scientific internationalism much as Berkner had outlined. They were soon disappointed, and indeed quite alarmed. Their optimism was misplaced for two reasons. The first was that Brode, whom they knew as a Washington insider (albeit not very well), actually shared Dulles's vision that international science needed to reflect national foreign policy aims. Like Dulles, Brode was convinced that ideology was the chief determinant of Soviet behavior, meaning that Soviet scientists were wholly subservient to the will of the state, and their findings of limited value. "Science is part of the Soviet ideology," Brode declared: "the Sino-Soviet Bloc has been and will continue to use science as a prestige, promotional and defense tool." US national security was harmed, in his view, by the reluctance of American scientists to show a united front to counter the Communist world's own use of science as a political and ideological weapon. Indeed, for Brode, few boundaries remained between military and non-military science, making it unwise for the US government to support international scientific meetings,

---

35 Detlev W. Bronk to Victor F. Weisskopf, [n.d.] [circa 1954], Rockefeller Archives Center, Sleepy Hollow, NY, Detlev W. Bronk papers (hereafter RAC, BRONK), Record Group 303.U, box 76, folder 2.

36 Detlev W. Bronk, "Consideration of present United States policy re membership of non-recognized regimes in the International Council of Scientific Unions (ICSU) and its associated unions," p. 3, [n.d.] [cover memo dated 23 November 1958], RAC, BRONK); see also Doel, Hoffmann, and Krementsov, "National States and International Science," p. 70.

"excepting where our political interests may profit by such."[37] To gain scientific intelligence from scientists outside the Western Bloc, Brode favored comparatively indirect means: the translation of Soviet and Chinese scientific periodicals (long practiced by the CIA), and the renewed placement of science attachés in US embassies located behind the Iron Curtain, with instructions to them to obtain information of strategic value.[38]

A second reason their optimism was misplaced was that Brode was not alone in his views within the US scientific community. Firm, vocal support for Department of State policies came from chemists, members of the largest professional scientific discipline in the United States, and an international powerhouse in chemical research. American chemists had won a large share of Nobel Prizes in chemistry since the start of World War II. In addition, chemists had been more involved in industry than members of any other discipline. Furthermore, the world's largest scientific organization, the American Chemical Society (ACS), had a strongly nationalistic orientation; the ACS charter required members to enhance "the development of US industries and [add] to the material prosperity and happiness of our people."[39] More politically conservative in general than their colleagues in physics and biology, many US chemists were quick to disparage research by Communist scientists, and many valued the ACS more than the International Union of Pure and Applied Chemistry (IUPAC), then one of the weaker and less cohesive of the international scientific unions. A leading US chemist in his own right, Brode felt comfortable with these views. "International scientific organizations," Brode confided in his diary, "are a minor factor of insignificant importance to the strength of world or national science and do not contribute appreciably to understanding between nations or peoples."[40] Buoyed by these beliefs, Brode, like Dulles, sought to make science a tool of US foreign policy, telling the astonished leaders of the US astronomical community—who had put forward the United States as host country for the proposed 1961 meeting of the International Astronomical Union, after the Soviet Union had become the IAU host in 1958—that Taiwan alone needed to represent China at the meeting

37 Wallace R. Brode, "International Scientific Cooperation," memo, 5 February 1960, p. 2, Library of Congress, Washington, DC, Wallace R. Brode papers, box 6 of 11.

38 Brode, "International Scientific Cooperation," and Wallace W. Atwood, Jr. and André C. Simonpietri to H.P. Robertson, strictly personal, California Institute of Technology archives, Pasadena, CA, H.P. Robertson papers, box 11, folder 2; I thank Simone Turchetti for providing me with a copy of this document.

39 Doel, Hoffmann, and Krementsov, "National States and International Science," p. 71.

40 Ibid., p. 72.

planned for Berkeley, and US delegates to the IAU needed to vote to exclude Communist Chinese scientists "on any technicality."[41]

By the late 1950s these deeply contrasting visions of scientific internationalism and secrecy came to a head, entirely in secret, at the highest levels of government. What created the opportunity to air these differences was Eisenhower's decision, after the launch of Sputnik, to create a new high-level advisory body, the President's Science Advisory Committee (PSAC). PSAC's leadership included Bronk, MIT President James Killian, and Manhattan Project leader George B. Kistiakowsky (political moderates all) as well as a few of the politically conservative scientists, such as Edward Teller, who had the ear of the President during his first term. In late 1959 Kistiakowsky, by then Eisenhower's science advisor, successfully proposed "international scientific cooperation" as an action item to be discussed by the National Security Council (NSC), the nation's highest policy-setting body. Kistiakowsky quickly discovered that Brode had supporters within the President's cabinet, and hence within the NSC. Commerce leaders backed NATO science as primarily military in character, while Elmer F. Bennett (Under Secretary of the Department of Interior) asserted that limits on international science were indeed justified: "as science increasingly affects international policy and scientists acquire more of a role in determining that policy," Bennett wrote, "it seems to me that some of the traditional freedom [of science] may be curtailed." In all, at least three Cabinet leaders sided with Brode, favoring a restricted definition of scientific internationalism fitting Cold War realities.[42]

Throughout 1960, Kistiakowsky strenuously promoted his own vision of scientific internationalism, arguing that supporting East-West contacts was an expression of democratic values, in sharp contrast to the view that defeating Communism required sacrifices consistent with total war. Both political traditions had been nurtured in early Cold War America, and were familiar to all involved: the key question for both sides was whether science was an exceptional activity that transcended US foreign policy. Over and over again, Kistiakowsky argued that scientific internationalism had eased political tensions, fostered 'evolutionary trends' favoring democratic values within Eastern

---

41 Ibid., p. 70.

42 Elmer F. Bennett to Detlev W. Bronk, 1 July 1959, RAC, BRONK, box 16, folder 15; E.B. Skolnikoff, "Substantive Comments on the Previous Draft of the [PSAC] Science and Foreign Affairs Panel Report," 16 July 1959, National Archives and Records Administration, College Park, MD, Records of the Executive Office of the President, Office of Science and Technology, Record Group 359, box 111; I thank Allan A. Needell for providing me with a copy of the latter document.

Bloc nations, aided scientific intelligence-gathering, and heightened US prestige worldwide ("Discussion with Soviet scientists," he declared, "whether in the USSR or elsewhere, is an effective way of obtaining useful technical intelligence, which often is not available in other ways because of their closed society.")[43] The arguments that Kistiakowsky and Brode advanced were both open and cloaked in secrecy. Kistiakowsky penned a spirited defense of scientific internationalism for publication in *Science*, while Brode crafted his 1960 Priestley Medal address to the American Chemical Society to promote his contrary views. Both publications were tips of far vaster icebergs, reflecting more explicit and entirely classified discussions behind closed doors in Washington in the run-up to the final NSC policy discussion.

In December 1960, that is, just weeks after the close election that made John F. Kennedy the 35th president of the United States, Eisenhower presided over the National Security Council meeting where US international scientific cooperation policy would be decided. After listening to competing arguments, Eisenhower decided in favor of Kistiakowsky's argument. NSC action item 2166-b-(15), a defeat for State Department conservatives, declared that "[i]nternational scientific activities relate directly and increasingly to the national security objectives of the US."[44] The decision marked a return to the arguments that Lloyd Berkner had articulated a decade before: that scientific internationalism and scientific intelligence gathering could be balanced in ways that allowed American scientists to serve their nation without violating perceived international standards.

Conclusion

In contrast to Germany under Nazi control, and the Soviet Union under Joseph Stalin, the United States government had not sought to place restrictions on international science in the 1930s. But by the early Cold War, scientific intelligence—and higher levels of secrecy—were increasingly of concern to American scientists, and to leaders of the nation state. For them, the Cold War fundamentally redefined the relationship between scientists and the state. National security, and national prestige, became pervasive influences in

43 Doel, Hoffmann, and Krementsov, "National States and International Science," p. 73; see also George B. Kistiakowsky, "International Scientific Activities Presentation to NSC, Dec. 1, 1960," classified secret, Dwight D. Eisenhower Presidential Library, Abilene, KS, Ann Whitman file, box 13.

44 Doel, Hoffmann, and Krementsov, "National States and International Science," p. 74.

decisions about international relationships in science as well as the freedom of scientists to meet with one another, to exchange data, and to cooperate in research programs. It challenged American scientists to rethink their long, deeply held conviction about the internationality of science and the duty of scientists to serve their country at a time when distinctions between military and civilian science became blurred.

It was not inevitable that the US government would back scientific internationalism as an aspect of national security policy in the late 1950s and early 1960s, rather than side with those who argued that international science needed to take a back seat to US foreign policy goals. That this decision took the form it did owed to several factors. One was the established international character of atomic energy (and of outer space, following the launch of Sputnik). A second, important reason, was the availability of patronage not controlled by the state for scientists (which, in this case, allowed the private Ford Foundation to assume the cost of maintaining US membership in international scientific unions when the Department of State balked at paying them). A final reason was the success scientists had in convincing White House leaders, including Eisenhower, that the Soviet Union was not a wholly closed society driven by ideological zeal, but rather a dictatorship in which scientists had limited but significant influence—an influence to the extent that providing aid to moderate scientists in the Soviet Union (and throughout the East Bloc) could bring about desired "evolutionary changes" and support democratization.[45] Despite their success in framing scientific internationalism as a positive tool for US national security, leading scientists also recognized that there was no clear consensus on this issue within the American scientific community, despite the near-unanimity among atomic physicists who supported this ideal without question. The willingness of many US chemists to support restrictions on scientific internationalism in order to give greater clarity to the nation's foreign policy objectives—and to abandon the framework of international scientific unions in favor of Western-led scientific entities, such as the American Chemical Society—revealed that certain scientific norms, rather than universal, were influenced by the perceived competitiveness of foreign scientific research. If the lines between civilian and military science were blurred, so too were the lines between scientific internationalism, elitism, and national ambitions.

It is prudent to remain skeptical that we now know how the conflict between scientific internationalism and intelligence gathering played out in Cold War

45 Ibid., p. 73.

America.[46] Too many key documents remain classified, even after the end of the Cold War.[47] Furthermore, our insights are largely restricted to discussions that occurred at the highest levels of government and leadership within the scientific community. What we still miss may be the most crucial: what it meant to provide scientific intelligence at lower levels. We know relatively little about scientists (the vast majority still unidentified) who served as analysts within the CIA's Office of Scientific Intelligence; we also have few insights into how the research of these individuals was integrated into national or foreign policy decisions—or indeed the extent to which these individuals circulated within the 'open' scientific community. Put another way: we do not yet know if what we glimpse here is genuinely a 'science in black'—an emerging covert research community, largely distinct from its overt academic counterpart—or a 'science in gray:' a community of researchers already well-known to their colleagues. It is one thing to know that the CIA's first scientific intelligence report OSI 1/49, "An Estimate of Swedish Capacities in Science," was largely accurate; it is quite another to know who actually wrote it, and what kinds of academic training and experience she, he, or they had. Such information would contribute significantly to illuminating how scientific intelligence gathering worked as a practice.[48]

We also remain in the dark about a still larger issue: during the early Cold War, how often did scientists employed in regular academic or professional posts moonlight as consultants, either occasionally or regularly, for the CIA and other intelligence agencies? A few cases have already come to light. For instance, the University of Chicago astrophysicist and planetary scientist Gerard P. Kuiper, anxious to determine whether a 1958 claim by a Soviet astrophysicist at Pulkovo Observatory was true (that the moon showed evidence

46 We also know relatively little about whether similar tensions played out in other nations; helpful exceptions include Paul Maddrell, "British-American Scientific Intelligence Collaboration during the Occupation of Germany," in David Stafford and Rhodri Jeffreys-Jones (eds.), *American-British-Canadian Intelligence Relations, 1939–2000* (London, 2000), 74–94; Paul Maddrell, *Spying on Science: Western Intelligence in Divided Germany, 1945–1961* (Oxford, 2006); and R.W. Home and Morris F. Low, "Postwar Scientific Intelligence Missions to Japan," *Isis* 84 (1993), 527–37; also helpful is Bruno J. Strasser and Frédéric Joye, "Une science 'neutre' dans la Guerre Froide? La Suisse et la cooperation scientifique européenne (1951–1969)," *Schweitzerische Zeitschrift für Geschichte* 55 (2005), 95–112.

47 For instance, at the time of writing, the entire collection for science advisor Jerome Wiesner at the John F. Kennedy Presidential Library remains closed and unavailable to researchers: see the online record for Jerome B. Wiesner Personal Papers at http://www.jfklibrary.org/Asset-Viewer/Archives/JBWPP.aspx [accessed 31 May 2014].

48 Doel, "Scientists as Policymakers, Advisors, and Intelligence Agents," p. 216.

of active volcanism), used CIA funds to hire a visiting Yugoslavian astronomer; this astronomer was then tasked with preparing a comprehensive review of Russian studies in planetary and stellar astrophysics (in part to assess the extent of ideological intrusions in that field). A naturalized US citizen who had worked in the Radio Research Laboratory at MIT during World War II and had participated in the ALSOS mission at the war's end, Kuiper was well-versed in the government secrecy practices; he kept the nature of his CIA contract secret from his colleagues at Chicago's Yerkes-MacDonald observatories, which he directed. Kuiper's correspondence is revealing: he advised his CIA handlers that the Soviet Lunik photographs of the moon's far side seemed real, rather than fakes, thus helping the US avoid an embarrassing charge of deception. Separately, he declared his unwillingness to deceive his Soviet colleagues for political gain, as he was apparently asked to do. Concern for national security partly explains Kuiper's willingness to work with the CIA. He also hoped to increase the competitive edge of his institution and his own research programs. But Kuiper's line-in-the-sand declaration that he would not deliberately share falsehoods with his Soviet colleagues suggests Mertonian norms in practice. Was this response common within the US scientific community? We cannot know at this stage, because there are too few case studies on which to judge.[49]

Perhaps the most important question concerns the circulation of knowledge. Historian James Secord rightly argues that one of the most important questions in the history of science is how knowledge flows: how it is carried from place to place, how it gains significance, how it moves from being local, private information to a level of global significance where its validity is 'taken for granted.'[50] In recent years we have learned how important individuals, even more than publications, are in communicating scientific information. Physicist J. Robert Oppenheimer may have expressed this best in declaring that "The best way to send information is to wrap it up in a person."[51] What did it mean when scientific information, wrapped up as scientific intelligence

49 Ronald E. Doel, "Evaluating Soviet Lunar Science in Cold War America," *Osiris* 7 (1992), 238–64.

50 James Secord, "Knowledge in Transit," *Isis* 95 (2004), 654–72; and Kapil Raj, "Beyond Postcolonialism … and Postpositivism: Circulation and the Global History of Science," *Isis* 104 (2013), 337–47.

51 David Kaiser, *Drawing Theories Apart: The Dispersion of Feynman Diagrams in Postwar Physics* (Chicago, 2005), p. 357; Tiffany C. Vance and Ronald E. Doel, "Graphical Methods and Cold War Scientific Practice: The Stommel Diagram's Intriguing Journey from the Physical to the Biological Environmental Sciences," *Historical Studies in the Natural Sciences* 40 (2010), 1–47; Abraham Pais, *J. Robert Oppenheimer: A Life* (New York, 2007), p. 90.

reports, flowed not through individuals up chains of state authority, but instead through official reports? What did it mean when university scientists who met at lunch, uncertain of the security classification held by their lunch-mate colleagues, censored the scientific ideas they felt able to discuss?[52] In 1956, the *Washington Post*, with the determined support of its then publisher Philip L. "Phil" Graham, argued that "[s]ecrecy in science has been carried to such lengths that it is now operating to stifle the national security it is supposed to protect."[53] But was scientific internationalism so tainted by military support for American research by then that the flow of scientific information had become limited in other, even more significant ways? ("[T]he majority of all our basic science programs are supported by military agencies as an altruistic gesture but with hidden motives," Brode had privately confided to handwritten file notes in 1959—a candid admission that reflected the very real worry of Western European scientists.) Just how did knowledge circulate, and gain credibility, in the kaleidoscopic Cold War? A portrait is emerging, but to flesh it out, we need insights from a great many fields.

---

52 "[Section] B. Scientific and Technical Information," p. 12, California State Library, Sacramento, CA, John E. Moss papers, box 371, folder 5.

53 "The Cult of Secrecy," *Washington Post*, 8 April 1956, E 4; see also Ronald E. Doel, oral history interview with Dr. Joseph B. Koepfli, 3 August 1995, Center for History of Physics, American Institute of Physics, online at http://www.aip.org/history/ohilist/31375.html [accessed 31 May 2014].

CHAPTER 2

A 'Need-To-Know-More' Criterion? Science and Information Security at NATO during the Cold War

*Simone Turchetti**

'Need-to-know' is one of the chief principles in use to protect classified knowledge. Its current definition describes it as "a determination made by an authorized holder of classified information that a prospective recipient requires access to specific classified information in order to perform or assist in a lawful and authorized governmental function."[1] The principle is associated with the doctrine of compartmentalization, and was pioneered during World War II. It originally aimed to make a government research environment less vulnerable to espionage, since those who were in a position to give away information on a project would not be able to reveal all its details in full.

Several security acts of the early 20th century refer to a lawful entitlement needed in order to access "information respecting to national defense."[2] But the prevention of access to the documentation available to individuals at their workplace typified the provisions of wartime projects in general, and the US atom bomb ('Manahattan') project more specifically.[3] During the post-war years this security principle featured in new national security legislation such as the US Executive Order 10290 of 1951, ruling that "within the executive branch," dissemination of information shall be limited to "persons whose

* Research for this paper was generously funded by the European Research Council (grant nr. 241009). I wish to thank the organizers of the conference "Cold War Science" at the Lorentz Center at Leiden University, the Netherlands (especially Corrie Kuster); Ineke Deserno and Nicholas Nguyen at the NATO Archive; Charlotte Erwin and her collaborators at the Caltech Archive; and James Peters at the University of Manchester Library Archive.

1 U.S.C. 50, p. 401 as of 4 January 2012 (also US Executive Order 12968), unofficial version available at www.law.cornell.edu/uscode/uscprint.html [accessed 18 March 2014].

2 The principle appears in the 1917 US Defense Act and the 1917 US Espionage Act. Both recall a similar ruling of the UK Official Secrets Act of 1911.

3 The project's director, General Leslie Groves, stated the importance of "making our people stick to their knitting." See Charles Thorpe, *Oppenheimer. The Tragic Intellect* (Chicago, 2007), p. 100. See also Brian Balmer, *Science and Secrecy. A Historical Sociology of Biological and Chemical Warfare* (Farnham, 2012), p. 8.

 | DOI 10.1163/9789004264229_003

official duties require knowledge of such information."[4] By then, short notices had appeared on the front flap of classified research reports in the USA and elsewhere, warning that those wishing to read them ought to have their 'need-to-know' certified by an authorized military agency.[5]

During the 1950s this security criterion underwent closer scrutiny, especially because of its implications for international collaborative research. With the 1953 Atoms for Peace program, the 1955 Geneva Conference on the Peaceful Uses of Atomic Energy, and the International Geophysical Year (1957–1958), officials in the US administration sought new mechanisms to share knowledge in order to boost collaborative projects. Traditional security provisions associated with compartmentalization now appeared obstructive; especially with regard to US ambitions to produce synergies in the world of science and building alliances through political and defense treaties. The shortcomings of the 'need-to-know' arrangement appeared evident, especially to those science administrators who intended to promote collaborative research in the context of the North Atlantic Treaty Organization (NATO). It prompted them to reconsider secrecy rulings, and pioneer what this article refers to as a 'need-to-know-more' criterion.

The management of secrecy has been the subject of many scholarly studies. Recent works have helped to reject the conventional reading of secret and open knowledge as two entirely independent frameworks of reference for scientific research. By analyzing patents,[6] regulations,[7] international meetings,[8] and the transfer of scientific knowledge across borders,[9] a new understanding of the constructed boundaries of secrecy and openness has emerged. Adjusting to existing security rulings was not always easy for those who sought to promote international collaborative work, as shown in Alan Needell's portrait of

4 US Executive Order in *Federal Register*, 27 September 1951, p. 9799.

5 For instance, see Henry A. Ossing, *Dispersion of Rayleigh Waves Along the Mid-Atlantic Ridge* (Bedford, MA, 1963), p. i.

6 Brian Balmer, "Secrecy as a Spatial–Epistemic Tool. A Secret Formula, a Rogue Patent and Public Knowledge about Nerve Gas," *Social Studies of Science* 36 (2006), 691–722; Simone Turchetti, "Patenting the Atom," in Stathis Arapostathis and Graham Dutfield (eds.), *Knowledge Management and Intellectual Property* (Cheltenham, 2013), pp. 216–34.

7 See, for instance, Alex Wellerstein, "A Tale of Openness and Secrecy: The Philadelphia Story," *Physics Today* 65 (2012), 47–53.

8 For example, see John Krige, "Atoms for Peace, Scientific Internationalism, and Scientific Intelligence," in John Krige and Kai-Henrik Barth (eds.), *Global Power Knowledge: Science and Technology in International Affairs* (Chicago, 2006), pp. 161–81.

9 This included the atom spy cases; see Simone Turchetti, *The Pontecorvo Affair: A Cold War Defection and Nuclear Physics* (Chicago, 2012).

US science administrator Lloyd V. Berkner.[10] And Paul Forman has revealed how those scientists who came to terms with Cold War security often faced its negative consequences (especially in terms of an inability to think "out of the box").[11]

The study of science and secrecy at NATO enriches our understanding of these issues, and explains why the reformation of information security was a decisive component in reconfiguring the production and circulation of knowledge during the Cold War. Although some historical literature on the defense alliance exists, which also covers its science activities, little has been written on its information security system.[12] Based on archival materials from NATO and other documentation centers, this article shows that security rulings adopted to administer scientific affairs helped to further assert US influence on the directions and priorities of new scientific programs sponsored by NATO. For this context, I suggest that secrecy worked as a key policy device in the administration of international scientific relations during the Cold War.

As the largest producer of classified studies in the alliance, and holder of its strategic nuclear deterrent, the US administration feared more than other nations the dissemination of scientific (and especially nuclear) information. Conversely, US efforts to strengthen relations within the alliance through science and technology might have failed due to opposition to sharing information. In order to strike a useful balance, a few US science administrators looked for a compromise, and sought ways in which to reform information security accordingly, with the assistance of trusted colleagues in Canada and Britain. Their solution was summarized in a NATO report and construed as an invitation to reconsider the application of the 'need-to-know' criterion. It also involved, as a token gesture, the release of classified scientific literature available only in the USA.

This solution highlights one of the chief characteristics of Cold War science: namely that the management of scientific affairs often replaced traditional

---

10 Alan Needell, *Science, Cold War and the American State: Lloyd V. Berkner and the Balance of Professional Ideals* (Washington, D.C., 2000), introduction and pp. 116–7.

11 "The separation of different pieces or classes of information within the mind of one and the same knower, the prevention of mental juxtaposition and cognitive integration of disparately labeled 'data'." Paul Forman, "Into Quantum Electronics: The Maser as 'Gadget' of Cold-War America," in P. Forman and José M. Sánchez Ron (eds.), *National Military Establishments and the Advancement of Science and Technology* (Dordrecht, 1996), p. 300.

12 See for instance Lawrence Kaplan, *NATO's First Fifty Years* (Westport, CT, 1999); Marc Trachtenberg, *A Constructed Peace. The Making of the European Settlement, 1945–1963* (Princeton, NJ, 1999). On science and technology sponsored by NATO see John Krige, *American Hegemony and the Postwar Reconstruction of Science in Europe* (Cambridge, MA, 2006).

diplomacy approaches to specific issues in a process of constructive evasion. Government officials agreed to dodge compelling issues in diplomatic talks, especially on nuclear weaponry, recognizing the lack of short-term solutions accommodating everyone's interests. But in order to keep diplomatic channels open, they put forward new proposals focusing on how to advance scientific collaborative work.[13]

In the case in point, a US proposition on security reformation allowed NATO national delegations the freedom to decide what scientific information to release, and what to keep under wraps and this was consistent with the model of partnership that US officials wished to introduce in the alliance at this point in time. The security proposals they set forward highlighted an orientation, since the knowledge sharing mechanism they agreed to propose lent support to a model of defense integration that was not based on what Marc Trachtenberg defines as an "independent, integrated European nuclear capability."[14] In essence, scientific information on the alliance's nuclear deterrent was not open for sharing with the allies, while these US officers agreed to exchange information that boosted coordination and integration of other scientific items, including research propelling air and sea defense. Their proposals thus favored the dissemination of 'need-to-know' literature on environmental knowledge, since its significance for defense operations and the surveillance of enemy forces was being increasingly recognized in the USA and elsewhere, while hampering that on the nuclear deterrent. In the next sections I analyse the shift in thinking about science and secrecy at NATO, and how the deriving information security provisions fell in line with a US agenda.

NATO and its Science Committee

In the words of former NATO Secretary General Manlio Brosio, the NATO Science Committee "was helped into being by the rumbling of rockets and the alarming 'bleeps' from the first Sputnik."[15] The Soviet satellite, launched on 4 October 1957, focused the attention of NATO planners on the need of

13 This presents similarities with the circumstances of the 1959 Antarctic Treaty. See Simone Turchetti, Simon Naylor, Katrina Dean and Martin Siegert, "On Thick Ice: Scientific Internationalism and Antarctic Affairs, 1957–1980," *History and Technology* 24 (2008), 351–76, on p. 359.

14 Trachtenberg, *A Constructed Peace*, p. 140.

15 Speeches Commemorating the 10th Anniversary of the Foundation of the Science Committee, 21 March 1968, p. 5, NATO Archives, NATO Headquarters, Brussels, Belgium (hereafter NATO Archives), AC/137-D/330.

investing more in research recognizing the existence of a technology 'gap' with the Soviet Union. It also made manifest the intention of promoting research that would strengthen the alliance's defense, thus setting forth for the first time the problem of what information could be made accessible to scientists sponsored by NATO.

Established in 1949, the North Atlantic Treaty Organization was composed of a civilian decision-making structure, the North Atlantic Council (NAC), and a Military Committee that was responsible (along with its Standing Group comprising of US, UK and France representatives only) for defense planning. From 1954 onwards, operational responsibilities were divided between two Supreme Allied Commanders (one for Europe, the SACEUR; and the other for the Atlantic, the SACLANT). In its early days the organization of NATO overlapped with parallel plans for a European Defense Community. The accession of the Federal Republic of Germany in 1954 made of NATO the only multilateral defense organization in Western Europe for the foreseeable future.[16] By then, a strategy of massive retaliation was in place. It dictated that the alliance would respond to a Soviet attack with all the weaponry available (including nuclear).[17]

NATO's strategic posture recognized the critical role played by science and technology in preparing for war. In 1959 the SACEUR, US General Lauris Norstad, contended that the enemy might outnumber Western forces, but NATO would prevail due to its "technical superiority." Thus, he argued, military commanders had "become more and more dependent on scientists for the development of more powerful and flexible weapon systems as a deterrent to Soviet aggression."[18] The request for mobilizing science at the service of massive retaliation eventually led to the establishment of defense research organizations, such as the Advisory Group for Aeronautical Research and Development (AGARD), and the office of SACEUR's Science Adviser at the Supreme Headquarters Allied Powers Europe (SHAPE). These organizations were devoted to applied studies, and flourished during the 1950s; they also established sub-groups and research laboratories. But when the AGARD chairman, Theodore von Kármán, proposed to transform the organization he led

16 Trachtenberg, *A Constructed Peace*, pp. 120–5.

17 Kaplan, *The Long Entanglement*, pp. 80–1. See also Peter Schneider, *The Evolution of NATO. The Alliance's Strategic Concept and its Predecessors, 1945–2000* (Munich, 2000). Information on NATO's past strategies can also be found in the NATO website www.nato.int/cps/en/natolive/topics_56626.htm [accessed on 6 February 2015].

18 L. Norstad's speech in Defence Research Directors, Summary Record of Meeting, 8 October 1959, Secret, NATO Archives, AC/137-(DR)R/2, pp. 3–5.

into a Scientific Advisory Board to serve the NAC and cover a broader set of research issues, his proposal was rejected.[19]

Conversely, from 1956 the NAC began discussions on sponsoring a program to stimulate integration within the alliance. In that year, it appointed "Three Wise Men" (the foreign ministers of Italy, Gaetano Martino; Norway, Harvard Lange; and Canada, Lester Person) to provide recommendations on what actions could help NATO to be further strengthened.[20] The experts argued that science and technology could help to forge economic cooperation as well as to improve the position of NATO nations in world affairs.[21] The report considered especially how "to increase the supply of scientists, engineers and technicians," as a means to improve NATO's economic synergies further.[22] Other NATO groups, including a recently appointed committee headed by US Senator Henry M. Jackson, and a task force chaired by the chemist Joseph Koepfli of the California Institute of Technology (Caltech), reached similar conclusions.

NATO discussions at this point were mainly informed by recent US assessments on the growing number of trained scientific personnel in the Soviet Union and catered for efforts to match this growth.[23] But the launch of Sputnik marked a shift in thinking about NATO's investment. On 16–19 December 1957 the NAC Ministerial Meeting met at the level of Heads of Governments. US President Eisenhower led a delegation that included MIT president James Killian as science adviser, and Koepfli as its chief consultant.[24] The meeting ended with the release of a declaration of principles reiterating the Three Wise Men's recommendations, but now placing additional emphasis on the military potential of an investment in a science program. Thus, the NAC Communiqué stated NATO's intention to set up scientific and technical cooperation in the

---

19 Jan van der Bliek (ed.), *AGARD. The History, 1952–1997* (Ilford, 1999), chapter 3, pp. 14–6.

20 On the Three Wise Men see Krige, *American Hegemony*, p. 201.

21 The Three Wise Men report is available at: www.nato.int/cps/en/natolive/official_texts_17481.htm [accessed 30 January 2014].

22 Ibid.

23 In particular Nicholas DeWitt's report *Soviet Professional Manpower*, officially sponsored by the National Science Foundation, put a 'spin' on funding the formation and recruitment of scientists in the West. David Kaiser has pointed out that DeWitt's research was completed at Harvard's Russian Research Center; a unit jointly sponsored by the US Air Force, the CIA and the Carnegie Foundation. See D. Kaiser, "The Physics of Spin: Sputnik Politics and American Physicists in the 1950s," *Social Research* 73 (2006), 1225–52, on p. 1228.

24 Report to the Council by the Task Force, 4 November 1957, Confidential, NATO Archives, C-M(57)130. On Killian see Joseph Koepfli's oral history interview by Ronald Doel, 3 August 1995, Montecito, CA, AIP Collection, available at [accessed on 6 February 2015].

recognition that "the full development of science and technology is essential to the culture, to the economy and to the political and *military strength* of the Atlantic community."[25] This anticipated the establishment of the Science Committee, which comprised of civilian scientists from member states. It was established to "speak authoritatively on scientific policy," and its chairman was to be the newly appointed Science Adviser to NATO's Secretary General.[26] But then the problem of how the committee could sponsor initiatives aiming at improving the alliance's "military strength" had not yet been solved. Would its members—civilian scientists—be provided with security clearances to know more about NATO defense research? Would the committee be in a position to distribute classified research reports while funding new multilateral research projects? And if so, would this distribution be based on the nationally agreed security classification, or a NATO-wide one? The first NATO Science Adviser addressed these issues immediately after the new committee was set up.

The Problem of Secrecy

When he arrived at NATO headquarters in Paris, the first task for the US physicist Norman Foster Ramsey Jr. was to consider the 'secrecy problem' and its implications for the activities of the Science Committee. For the evaluation of new information security policies the newly appointed NATO Science Adviser could call on the assistance of two other science administrators that were well-known to him as the US delegates at the Science Committee: Isaac Isidor Rabi and Howard P. Robertson. These science administrators understood the important role that information security was going to play in the new committee's activities; they also had extensive experience with similar issues in managing (open and classified) research both in the USA and at NATO.

Howard P. Robertson was one of the founding fathers of operations research in America. In the 1930s he had risen through the academic ranks becoming Assistant, Associate, and then full Professor of Mathematical Physics at Princeton University. There, in the vicinity of Albert Einstein, he elaborated a cosmological model in collaboration with the British mathematician Arthur Geoffrey Walker (the so-called Robertson-Walker metric) consistent with the

25 Emphasis not original. Declaration and communiqué issued at the NAC Ministerial Meeting. Reprinted in *NATO: Facts About the North Atlantic Treaty Organization* (Utrecht, 1962), pp. 279–84, on p. 282.

26 Ibid., p. 283.

theory of relativity.[27] But during and after the war Robertson became more interested in the management of scientific affairs in national and international arenas. In 1945 he was appointed as Chief of the Scientific Intelligence Advisory Section of the Allied Forces Supreme Headquarters in Europe, and was involved in the planning of military activities in North Africa together with General Norstad.[28] The American mathematician's interest for operations research grew more intense, while that for the cosmological foundations of Einstein's relativity waned, and Norstad instigated his appointment as SACEUR's first science adviser in 1954. Although Robertson returned to the US two years later, he continued to play a key role in the administration of NATO scientific affairs. He also retained advisory roles in the CIA, the National Academy of Sciences, the Department of Defense, the National Security Agency and the RAND Corporation.[29] Presumably few knew better than Robertson how to navigate in the uncharted waters of information security: Robertson was constantly required to balance, on one hand, the advantages to US diplomatic work that were to be derived from releasing more information, and on the other, the need to keep critical details of national scientific programs hidden.[30]

Physicist Isaac Isidor Rabi, the other US delegate at the NATO Science Committee was, like Robertson, very familiar with the administration of security issues. By the time he was appointed to lead the US delegation, he had already dealt many times with the problems associated with the compartimentalization of research imposed by defense administrations in the US. He believed that these security rulings had a negative impact on scientists' performance, as they hampered the creative process by curtailing access to key pieces of information and limited the dialogue between researchers. A PhD graduate of Columbia University, he pioneered radar research at the MIT Radiation Laboratory during the war. At the end of the conflict he left MIT to return to Columbia (and chair its physics department), and was given positions of authority in a number of government advisory committees. In 1949, as member of the US Atomic Energy Commission's General Advisory

27 Jesse L. Greenstein, "Howard Percy Robertson, 1903–1961," *National Academy of Sciences. Biographical Memoirs* 51 (1980), 343–64, on p. 346.

28 Reginald Victor Jones, *Most Secret War. British Scientific Intelligence, 1939–1945* (London, 1978), pp. 480–2. See also Solly Zuckerman, *Men, Monkeys and Missiles: An Autobiography, 1946–1988* (London, 1988).

29 Pamphlet produced on occasion of Howard P. Robertson's death in 1962. Copy in Archive, California Institute of Technology, Pasadena, CA, USA (hereafter Caltech), Lee DuBridge Papers. See also Greenstein, "Howard Percy Robertson."

30 As discussed in Howard P. Robertson, "The Impact of Science on Military Thought," H.P. Robertson Papers, Caltech, box 18, folder 12.

Committee, Rabi famously wrote, together with Enrico Fermi, the "minority report" opposing the setting up of a hydrogen bomb program.[31] In the post-war years Rabi developed a new understanding of the importance of the cooperation between civilian and military scientists for the innovation of defense. It was particularly because of his awareness of the problems that compartmentalization presented that he sought to limit its negative effects on academic life when he returned to Columbia University. Rabi was appointed head of the project for the declassification of reports on wartime radar research; and this would enable the use of radar apparatus in radio-astronomy and radio-meteorology. He also used military funds to establish the Radiation Laboratory at Columbia University. As Forman has noted, Rabi kept the laboratory's technical program "intellectually compartmentalized, academically ostracized" to forestall attempts by military authorities to change its working methods and adopt tighter security provisions.[32] Rabi was highly esteemed in Europe, especially since he had worked towards making Marshall Plan funds available for the recovery of European laboratories, and had helped establishing CERN.[33]

Together, Rabi and Robertson outlined proposals on information security in order to assist Ramsey, who was a most esteemed colleague to both. Harvard physics professor Ramsey had successfully contributed to the program that Rabi had set up during WWII and was one of his assistants in wartime radar research.[34] In the post-war years Robertson and Ramsey had both been members of the same influential committees, such as the USAF Science Advisory Board, and in 1954 Robertson had warmly supported Ramsey's appointment to the National Security Agency—albeit without success.[35] There is no definitive evidence that Robertson was directly involved in the sequence of events that led to Ramsey's appointment as Science Adviser, but it was Robertson who

31 Peter Galison and Barton Bernstein, "In Any Light: Scientists and the Decision to Build the Superbomb," *Historical Studies in the Physical and Biological Sciences* 19 (1989), 267–347 and Paul Forman, "Inventing the Maser in Postwar America," *Osiris* 7 (1992), 105–34, on p. 106.

32 Forman, "Into Quantum Electronics," p. 301.

33 As noticed by Krige, Rabi was instrumental in overcoming "the centrifugal force of nationalism" in Europe. See John Krige, "Isidor I. Rabi and the Birth of CERN," *Physics in Perspective* 7 (2005), 150–64, on p. 162. On Rabi see also John S. Rigden, *Rabi: Scientist and Citizen* (New York, 1987).

34 Ramsey also worked on the magnetron device and analysed the physical parameters of bombers capable of delivering atomic weapons. See David Wineland, "Norman Ramsey (1915–2011)," *Nature* 480 (2011), 182.

35 A.B. Clark, NSA, to Robertson, 11 February 1954, H.P. Robertson Papers, Caltech, box 18, folder 13.

inducted the new adviser to NATO's workings as an organization. In April 1958, for instance, he invited Ramsey for a briefing at the Pentagon and argued for the importance of keeping the new committee's agenda in line with existing defense programs. Science and technology, Robertson stressed, ought to be utilized to assist planners to "evolve strategic and tactical plans and formulate military requirements for weapons in support of those plans."[36]

Thanks to Robertson's advice Ramsey realized that secrecy was a serious problem for the new committee. His first initiative was to put a list of questions before national delegations during the first Science Committee meeting, which was scheduled to take place on 26–28 March 1958. Ramsey's questionnaire focused on information security and asked if the committee should meet in "closed executive sessions," and how relevant research information could effectively be released by (and to) defense research agencies. One question featured more than once in the document: "how can the secrecy problem be overcome?"[37]

When the Science Committee met for the first time, answers to the questionnaire informed the discussion. Thirteen national delegates, most of them heads of a national research council (or science advisors to their governments), analysed existing information security rules and their implications for the activities of the new committee. Aside from Ramsey and Rabi, another authoritative voice was heard in these talks: that of the British delegate Solly Zuckerman. Chief Scientific Adviser to the UK Ministry of Defense from 1959, and to the British Government from 1964, Zuckerman was one of Robertson's friends and introduced him to Norstad during World War II because of their respective duties in planning military operations.[38] Since Zuckerman was much more friendly with the US representatives than the other delegates, it is not surprising that he provided support to their proposals on various occasions. NATO military authorities were also invited to the meeting, and because their presence raised the issue of what security measures should be in place to avoid unwanted spread of classified information, delegates unanimously agreed that the Science Adviser should be in charge of decisions on whether

---

36 Robertson to Buck Weaver, Department of Defense, 7 April 1958, H.P. Robertson Papers, Caltech, box 18, folder 5.

37 Suggested Subjects for Discussion and Action by the Science Committee, Appendix 1, Restricted, in Science Committee, Summary Record of Meeting held on 25 April 1958, NATO Archives, AC/137-R/1, pp. 46–9.

38 Zuckerman, *Men, Monkeys and Missiles*, pp. 155–9, 172–3.

sessions were open and closed, on a case-by-case basis.[39] Yet, information security issues continued to feature in the proceedings. Rabi noted that some Science Committee documents were labeled 'NATO restricted' and, since this classification marking corresponded to 'confidential' in the USA, he was prevented by law from discussing their content with unaccredited US officials. For this reason the committee agreed to request that the NAC reduce the classification for committee documents as much as possible.[40]

The national delegates next debated at length whether or not the Science Committee ought to commit new research to defense goals, and if so, how to expedite the dissemination of classified information. The Belgian, Canadian, British and French representatives all supported the viewpoint (to varying degrees) that the committee should promote research of use to the NATO defense establishment. Military authorities concurred. In particular, the Standing Group representative argued that, notwithstanding the 'Three Wise Men's recommendations, the committee's activities "should be additional to, and not at the expense of, the military effort." The basis of the Alliance was "the need for collective defense" and the military was "the leading customer for the end product of scientific R&D."[41]

Rabi intervened and stressed the need to avoid making decisions on "extremely delicate" matters before the Science Adviser conferred with "the appropriate agencies and individuals."[42] In this way he provided Ramsey with the opportunity to analyse how to combine the US interest in sharing some information and knowledge on defense with the allies with the impediments deriving from information security. Zuckerman supported Rabi's request that matters be left in Ramsey's hands. Robertson's friend from Britain made it clear that, although the Committee should address defense matters, it was the Science Adviser who should make decisions about what exactly ought to be discussed.[43] Rabi "strongly supported" Zuckerman's proposal, and the Committee deliberated that Ramsey would handle the security problem before the next

---

39 Science Committee, Summary Record of the Meeting held on 25 April 1958, NATO Archives, AC/137-R/1, p. 5.

40 Ibid., p. 6.

41 Major General Theodore W. Parker, Invited Statement on Matters of Defence Science, Secret, 6 May 1958, NATO Archives, AC/137-D/10.

42 Ibid., p. 30.

43 Zuckerman's support was in line with Britain's position, since a recently outlined Mutual Defense Agreement was about to allow Britain to be the only country in Europe to share information on the nuclear deterrent with the USA.

meeting.[44] In the meantime—as the following section will show—Ramsey, Rabi and Robertson were able to take full advantage of a proposal that the Canadian representative at the Science Committee put forward.

The Steacie Criterion

Edgar William Richard ('Ned') Steacie emerged as an authoritative figure in the Science Committee proceedings during the first meeting, but his prominence would not last for long. Indeed, that first meeting was one of the few he would attend at NATO headquarters. He fell ill shortly after, was diagnosed with cancer and passed away three years later.[45] A propagandist for declassification in the discussions that took place during the meeting, Steacie urged Ramsey to see what could be done for an increase in the circulation of knowledge. This, he argued, would entail "breaking down security barriers and ensuring that there was a proper exchange of information without which there could be no advance." He also claimed the 'need-to-know' to be "a major folly since it prevented the interchange of ideas between the various branches of research."[46]

Steacie's past and present activities as science administrator helped to foster these views. A physical chemist born in Montreal, Steacie was appointed Deputy Director of the wartime Anglo-Canadian atomic energy project Tube Alloy in 1944. He was thus, like Rabi and Ramsey, privy to the mechanics of wartime compartmentalization, but also aware of its limitations; according to one biographer, "he observed the destructive nature of secrecy enveloping these scientific efforts and its effects on the morale of the workers."[47] At the end of the war he was nominated director of the Chemistry Division at the Canadian National Research Council, and eventually became the council's new president. Because of his new role, he became involved in the work of agencies devoted to international cooperation at the end of the 1950s. This led him to condemn the effects of secrecy on research even more ferociously. In a speech at the 1957 meeting of the International Union of Geodesy and Geophysics he stated that

---

44 Science Committee, Summary Record of the Meeting held on 25 April 1958, Restricted, NATO Archives, AC/137-R/1, p. 31.

45 Léo Marion, "Edgar William Richard Steacie," *Biographical Memoirs of the Fellows of the Royal Society* 10 (1964), 257–81, on p. 263.

46 Science Committee, Summary Record of the Meeting held on 25 April 1958, NATO Archives, AC/137-R/1, pp. 29, 36.

47 M. Christine King, *E. W. R. Steacie and Science in Canada* (Toronto, 1989), p. 85.

"secrecy means nationalism of the most extreme form."[48] His views certainly made him popular with those officials in the US administration seeking to promote international scientific relations. In the same year he was elected foreign associate of the US National Academy of Sciences (NAS); of which Robertson was now the foreign secretary. Steacie also represented Canada at UNESCO's International Advisory Committee on Research in the Natural Sciences, where he made the acquaintance of the NAS director Wallace W. Atwood.[49]

The Canadian's bold claims certainly pleased the newly-appointed NATO Science Adviser Norman Ramsey, as well as the other administrators involved in US scientific relations. Ramsey now proposed that Rabi and Steacie establish "a small working party [...] to examine ways and means of bringing about large scale de-classification of scientific information."[50] Steacie worked on the task with proficiency, and on 15 May 1958 the draft "Scientific Secrecy and the NATO Science Committee" was ready.[51] The six-page *précis* argued that the committee "must see that secrecy does not unnecessarily obstruct scientific advance," and that those responsible for classification should distinguish between "major scientific goals" deriving from disclosure and "marginal military advantages deriving from enforcing classification." Secrecy should thus apply only when "military advantage outweighs the impetus to scientific progress that full disclosure would give."[52] The science committee should thus "exert its influence to ensure that the 'need-to-know' criterion is not allowed to obstruct the exchange of scientific and technical information within the NATO community." A balance should be sought between "security through secrecy" and "security through achievement."[53] National security would be better protected through new scientific accomplishments that gave a military advantage rather than by the curtailment of scientific information. Steacie's rhetoric patently echoed that of one of the most influential science policy documents in the post-WW2 era: *Science: The Endless Frontier*. In this 1945 publication, Vannevar Bush, the former head of the US Office of Scientific Research and Development

48 Ibid., p. 179. See also Léo Marion, "Edgar William Richard Steacie," pp. 259–62.

49 Members of the International Advisory Committee on Research in the Natural Sciences, 22 October 1958, available at: http://unesdoc.unesco.org/images/0012/001267/126707eb.pdf [accessed on 5 February 2015]. See also King, *E. W. R. Steacie and Science in Canada*, p. 174.

50 Science Committee, Summary Record of the Meeting held on 25 April 1958, NATO Archives, AC/137-R/1, p. 37.

51 E.W.R. Steacie, Scientific Secrecy and the NATO Science Committee, Ottawa, NRC, 15 May 1958, 7 pages. Copy in H.P. Robertson Papers, Caltech, box 18, folder 4.

52 Ibid., p. 3.

53 Ibid., pp. 5–6.

(also responsible for the Manhattan Project), had argued that scientific progress, rather than classification, was essential to national security, and that tight compartmentalization held back scientific knowledge.[54]

The resonance of Steacie's document with Bush's pioneering science policy manifesto certainly helped officials in Washington DC to consider the merits of Steacie's draft. On 12 June 1958 the document was discussed by the so-called 'backstopping group.' This was the US committee in charge of formulating policy positions to be endorsed by US delegates at the Science Committee; it included Robertson, Rabi, and representatives from the Department of Defense, the National Science Foundation, State Department's International Cooperation Administration, the AEC, and the White House. While it is not entirely clear what the US officials discussed, they certainly weighted the advantages to be derived from information sharing with the allies with the risks to national security of extending the sharing mechanism to items vital to the US nuclear defense. Even after the Berlin crisis, Eisenhower's administration had carefully avoided discussing the possibility of sharing knowledge on nuclear weapons with its European allies (excluding Britain); something that made some governments more eager to pursue independent nuclear projects either alone (France) or in collaboration with others (Italy, France and West Germany).[55]

During the meeting the backstopping group failed to reach consensus on whether or not to endorse Steacie's document.[56] However, five days later Robertson wrote to Steacie, stressing that both Rabi and he agreed with the proposal; they just wanted to amend the recommendations in order to state US policy more clearly with regard to the dissemination of scientific information at NATO. In fact, Robertson and Rabi now wished to use the document to take forward a US agenda on the NATO science program. They thus added one absolutely decisive amendment, stating that dissemination would not cover scientific information with "an active and direct military" significance.[57] This

54 Vannevar Bush, "Science: The Endless Frontier," NSF Report, 1945, available at http://www.nsf.gov/od/lpa/nsf50/vbush1945.htm [accessed on 3 February 2015]. Notably, Steacie wished to establish a Canadian NSF; see King, *E. W. R. Steacie and Science in Canada*, p. 142.

55 On Eisenhower's opposition see Trachtenberg, *A Constructed Peace*, pp. 274–282. On the ambitions of France, West Germany and Italy see Jeffrey T. Richelson, *Spying on the Bomb*, New York: W. W. Norton & Co., 2007, pp. 205–213. See also Paolo Cacace, *L'Atomica Europea* (Rome: Fazi, 2004).

56 H.P. Robertson to Ned Steacie, 13 June 1958, H.P. Robertson Papers, Caltech, box 18, folder 4.

57 Robertson to Steacie, 17 June 1958, H.P. Robertson Papers, Caltech, box 18, folder 4.

essentially enforced the withholding of knowledge on two items in particular: information on nuclear weapons and the means to deliver them. Critically, however, by endorsing Steacie's document they manifested their intentions to include in the sharing mechanism all the other items that would strengthen it, especially knowledge on scientific means to facilitate integration of air and sea defense, and the surveillance of enemy forces. Apart from the Eisenhower administration's policy on defense assistance to Europe, the decision was also informed by new guidelines on classification, which the US president was about to introduce to facilitate international talks. His Science Advisory Committee, including both Rabi and Robertson, met at the White House the day after the draft was amended. At that meeting a discussion ensued on the forthcoming Geneva negotiations of a ban of atomic weapons testing. On that occasion the US President and his advisers agreed to declassify information on seismic detection of Soviet tests, in order to encourage a positive outcome of the international talks.[58]

Robertson and Rabi also endorsed Steacie's report because it combined bold and progressive stances on secrecy with the lack of specific instructions. Ultimately the document represented a declaration of intent, leaving decisions on the documents to be released by national delegations to their staff. This left US research administrators free to decide what they deemed suitable for international collaboration with their Western allies. The document was made available to the Science Committee on 4 July 1958, and by then Rabi and Robertson had already agreed on additional measures that would further impress NATO allies and convince them to enter the new information security regime that they wished to promote.

Access to ASTIA

During his presentation at the third Science Committee meeting, Ramsey stated that its members "should impress upon their governments the need to increase the free dissemination of scientific information, particularly of

---

58 See Memorandum of Conference with President Eisenhower, 18 June 1958, FRUS. Available at http://history.state.gov/historicaldocuments/frus1958-60v03/d159 [accessed on 5 February 2015]. This position was, however, short-lived; see Simone Turchetti, "In God We Trust, All Others We Monitor: Seismology, Surveillance and the Geneva Test Ban Negotiations," in Simone Turchetti and Peder Roberts (eds.), *The Surveillance Imperative. The Rise of the Geosciences during the Cold War* (London and New York: Palgrave, 2014).

classified information in the defense research field."[59] At that stage, two important steps had been taken to reform information security in the way he advocated. A mechanism by which classified information could be made available to, and discussed by, Science Committee members was now agreed upon. And the US delegation divulged a new scheme for classified information produced in the USA, to be made immediately available to its allies.

Rabi's plan to introduce a mechanism of information-sharing at NATO included the release of abstract bulletins to the committee, as well as other classified technical documentation available to the US Armed Forces. Rabi had already indicated his intention to explore this possibility during the first Science Committee meeting, when he referred to the 'need-to-know' literature routinely collected by the US Armed Forces Technical Information Agency (ASTIA).[60] Back in the spring of 1958, Rabi and Robertson had forwarded a request to the US Deputy Assistant Secretary of Defense for Research and Engineering, John B. Macauley. On 12 May Robertson was informed that, in order "to carry out the US commitment concerning scientific cooperation [at NATO] and to support Dr. Ramsey in his official capacity," ASTIA would prepare a special bulletin containing publications labeled "confidential" for the NATO Science Adviser. Additionally, the Science Adviser was allowed to request classified publications listed in the bulletin. Secret publications could be made available too, but only on a "non-releasable to foreign nationals (NOFORN)" basis.[61]

The decision enabled the Science Committee members to gain access to what was the largest library of classified scientific knowledge in the USA at the time. The ASTIA, established in 1951, succeeded the US Air Force Central Documents Office, and both agencies inherited their functions from a wartime allied scientific intelligence agency based in London.[62] ASTIA functioned as central library of all technical information available to the US Navy, Army and Air Force, and all organizations devoted to defense research used it as repository. Consequently the collection grew at a pace of thirty thousand reports per

---

59 Science Committee, Summary Record of the Meeting held on 19 January 1959, Secret, NATO Archives, AC/137-R/3, pp. 25–6.

60 Science Committee, Summary Record of the Meeting held on 25 April 1958, NATO Archives, AC/137-R/1, p. 37.

61 Paul D. Foote, Memorandum for the Secretaries of the Army, Navy and Air Force, Official Use Only, 12 May 1958. Copy in H.P. Robertson Papers, Caltech, box 18, folder 4.

62 Renamed in 1963 'Defence Documentation Center,' and in 1979 'Defense Technical Information Center,' the agency is still active today. See "DTIC History" at http://www.dtic.mil/dtic/pdf/aboutus/dtichistory.pdf [accessed 13 December 2013].

year. By 1961 it consisted of approximately 250.000 papers. It is worth noticing that the high volume of technical information processed led ASTIA to pioneer automated methods of information search and retrieval.[63] By making available the ASTIA titles to NATO, Rabi and Robertson quite evidently hoped to energize the committee's activities and materialize Steacie's concept that NATO security depended upon achievement, and not just security.

During the third committee meeting new regulations were approved that marked the attempt to go beyond 'need-to-know' provisions. Rabi stated that the release of the ASTIA bulletin aimed to stimulate essential defense research, and Zuckerman followed suit claiming that the UK would also release classified documents in the future. Although the committee did not entirely agree, it deliberated that, "subject to certain reservations on the 'need to know,' the liberalization of the exchange of technical documentation [...] was extremely important and desirable."[64] The committee eventually approved a revised version of Steacie's report, which also made it possible for its members to discuss defense-related issues.[65]

Meanwhile, Ramsey had followed Zuckerman's suggestion that he alone investigate the selection of defense problems that should receive priority. During the fall of 1958 he organized a meeting of national representatives of defense research establishments (or Defense Research Directors, DRD) who met at NATO headquarters in Paris on 1–2 December 1958. A list of subjects of military importance with "a large element of scientific research" was distributed to the attendees in advance, in order to stimulate a discussion.[66]

This was the moment when the US delegation present at the meeting, of which Robertson was a member, capitalized on statements indicating that

63 José-Marie Griffiths and Donald W. King, "US Information Retrieval System Evolution and Evaluation (1945–1975)," *IEEE Annals of the History of Computing* 24 (2002), 35–55, on p. 42; Don R. Swanson, "Historical Note: Information Retrieval and the Future of an Illusion," *Journal of the American Society for Information Science* 39 (1988), 92–8.

64 Science Committee, Summary Record of the Meeting held on 19 January 1959, Secret, NATO AC/137-R/3, pp. 25–6.

65 Ibid., p. 27, and Science Committee, Report by the Working Party on Declassification of Scientific Information, Restricted, 4 July 1958, NATO Archives, AC/137-D/17.

66 DRD, Subjects of Military Importance Which Have a Large Element of Scientific Research, Restricted, 28 November 1958, NATO Archives, AC/137(DR)-WP/1. The NATO Secretary General, Paul-Henri Spaak, stressed that the DRD was not meant to be a new NATO committee, but just an informal forum assisting the Science Adviser in considering research priorities. However, it eventually became the main policy body responsible for NATO defense research. DRD, Summary Record of the Meeting held on 21 January 1959, Confidential, NATO Archives, AC/137(DR)-R/1, p. 4.

NATO should concentrate on research that would boost defense integration. Responding to Ramsey's request to clarify on areas which the Science Committee ought to prioritize, the DRDs of Canada and West Germany emphasized the importance of studies on underwater acoustics, the propagation of signals in the ionosphere, and geophysics.[67] The Norwegian delegate, Finn Lied, explicitly mentioned the need to "investigate all our geophysical environments in the widest sense and from a military point of view."[68] Robertson and Macauley, representing the US, recalled the lack of knowledge on the characteristics of the upper atmosphere, and they referred to the importance of geophysical studies for navigation and communication.[69] Their talk instigated reflection on research priorities and envisioned the urgency of a NATO research program focusing on the study of the "defense environment" (namely the physical space in which defense operations are planned). It thereby reaffirmed a position already stated by US and NATO military authorities—but known only to Ramsey. For instance, a 'cosmic top secret' SACLANT statement, which had been prepared for the second Science Committee meeting, emphasized the importance of oceanographic research to assist NATO anti-submarine warfare plans.[70]

Because of the decisions taken on secrecy by the Science Committee, related information could now be disseminated more easily, and at Ramsey's discretion, to its members. Ramsey made available to delegates a summary of items discussed during the DRD meeting, indicating the areas which the defense research directors hoped to see prioritized.[71] After a lengthy discussion, the committee agreed to call in groups of experts, who would envision and realize a program of collaborative investigations in the fields of oceanography and meteorology.[72] In the spring of 1959, when Frederick Seitz, the American pioneer of solid state physics, replaced Ramsey as Science Adviser, he announced the beginning of a NATO research grants program. This took the plans, for which Ramsey had already provided much of the groundwork, to completion.[73]

---

67 Ibid., pp. 14, 23.

68 Ibid., p. 38.

69 Ibid., pp. 50–4.

70 Science Committee, Statement on future ASW trends, Cosmic Top Secret, 17 July 1958, NATO Archives, AC/137-D/20. For an analysis of its implications see Simone Turchetti, "Swords, Shield and Buoys. A History of the NATO Sub-Committee on Oceanographic Research, 1959–1973," *Centaurus* 54 (2012), 205–31.

71 Annex A to Science Committee, Summary Record of the Meeting held on 19 January 1959, Restricted, NATO Archives, AC/137-R/3, pp. 25–6, 46–7.

72 Ibid., pp. 42–3.

73 Ibid., p. 23.

NATO Environmental Studies and Surveillance

From 1960 onwards the Science Committee funded fellowships, scientific meetings (the so-called Advanced Study Institutes) and research projects at the level of four million dollars per year. While the rationale for the first two strands of funding was to expand scientific manpower and encourage scientific exchanges, the NATO research program primarily aimed to boost cooperation in areas of fundamental research that were tied to defense matters. By 1965, five Science Committee sub-groups had been established to focus on specific research problems. Three of them dealt with environmental research on oceanography, meteorology and radio-meteorology, in agreement with the recommendations made by the DRDs.[74] The new information security provisions played a key role in the setting up of these programs, as they created the possibility to share scientific literature previously available only to US defense scientists.

One important example that shows advantages to be derived from the availability of ASTIA documents is a project for the study of propagation characteristics of signals from satellites; this project was funded from 1962 to 1965, and received the largest grant in the first five years of existence of a NATO research program ($154.000). The funding allowed European centers to study the propagation of signals from orbiting satellites in the upper atmosphere, and was coordinated by Nello Carrara at the Centro Microonde of Florence (Italy).[75] The project brought together Carrara's unit with the Jodrell Bank Observatory of Manchester (Britain); the Laboratoire de Physique de l'Atmosphere, Paris (France); the Ionosphären-Institut of Breisach (Germany); the Ionospheric Institute of Athens (Greece); and the Norwegian Defense Research Establishment center in Lillestrøm.

The project's instigator was Jules Aarons, the head of the Propagation Sciences Division at the Geophysics Research Directorate of the US Air Force Cambridge Research Laboratory.[76] Aarons helped establish relations with European groups from 1954 onwards. Previously, he had completed a doctor-

74 The scope of this paper precludes an analysis of the other two groups whose studies on operations research and human factors tied to defense needs.

75 Nello Carrara, Research Proposal on Propagation Characteristics with Satellites, NATO Surveillance Satellite, September 1961, Jodrell Bank Archive, John Rylands University Library, Manchester, UK (hereafter JBA), folder CS7/38/1.

76 Sunanda Basu, "In Memoriam: Jules Aarons," *IEEE Antennas and Propagation Magazine* 51 (2009), 137–8.

ate with Etienne Vassy, head of the Paris-based laboratory for atmospheric physics, and then worked together with AGARD and NATO scientists, inviting them to the USA to train and collaborate in specific projects.[77] When Sputnik began to send its 'bleeps,' Aarons was instrumental in suggesting research funding for Jodrell Bank and Carrara's institute in Florence; the funding was intended to assist with tracking down the satellite.[78] When more Soviet satellites began orbiting around the earth, he considered the potential of a multilateral effort under the auspices of the NATO, uniting research centers in the US and Western Europe. The expansion of collaboration to research groups of other NATO countries derived from the understanding that a comparative analysis on the propagation of signals, collected at different latitudes, offered invaluable 'surveillance-related' information. In particular, the NATO project aimed to use data and models on the ionospheric trails left by Soviet satellites as a proxy to establish their position and trajectory.[79]

Aarons and the other scientists operating in the USA routinely filed documents discussing their techniques and results in propagation studies as 'need-to-know' ASTIA reports. Carrara had also filed ASTIA reports while working under USAF contracts.[80] Under these circumstances, the existence of 'need-to-know' provisions may have hindered the collaborative work that the NATO Science Committee wished to sponsor. Conversely, the existence of provisions which facilitated the distribution of ASTIA 'need-to-know' documents amongst allies catered for open and regular exchanges of information between NATO research groups, especially those which were interested in the study of the ionosphere. The adoption of these new provisions thus allowed international collaboration at NATO to develop further, above and beyond traditional information security rulings.

---

77 Milton Greenberg to Von Karman, 26 October 1956, Theodore Von Karman papers, Caltech, box 34, folder 18.

78 See Jon Agar, "Moon Relay Experiments at Jodrell Bank," in Andrew J. Butrica (ed.), *Beyond the Ionosphere. Fifty Years of Satellite Communication* (Washington, D.C., 1997), pp. 19–30, on p. 23. See also Jon Agar, *Science and Spectacle. The World of Jodrell Bank in Postwar British Culture* (London, 1998); Franco Samoggia, *Nello Carrara* (Florence, 2006).

79 Nello Carrara, Research Proposal on Propagation Characteristics with Satellites, NATO Surveillance Satellite, September 1961, *JBA*, folder CS7/38/1.

80 Nello Carrara and P.F. Checcacci, Radar-Echo Rate from Meteors. Part 1: Theoretical Investigation, 10 April 1960, USAF Cambridge Research Laboratory, ASTIA AD0284504.

Conclusions

Although information security provisions associated with compartmentalization were consistently implemented at a national level during WWII, it was mainly during the Cold War that they found application in the activities of international agencies. Because of the intensification of the Cold War, globally, the design of new strategies for the production and circulation of knowledge across and beyond national borders played a prominent role in shaping Cold War Science. Important decisions needed to be taken on the type of information that could be disseminated, especially in light of considerations on how the progress of science and technology would impact on military thinking and operations. Moreover, secrecy—at NATO and elsewhere—worked as a device for the reconfiguration of research ambitions and goals. Information security helped to set new trajectories in individual institutions by blocking access to some research items, and removing obstructions to others.

This paper has shown that decisions about information security occupied a prominent space in the proceedings of the NATO Science Committee. The new science policy group was devoted exclusively to sponsoring the production of (unclassified) fundamental knowledge in the context of a new research program; this program had been decided in accordance with criteria set out by the NATO Science Adviser and the other committee members. Yet, by the time the committee began to operate, the NATO delegates agreed to align the committee's research agenda to defense priorities. As a consequence, the implementation of plans that would put new research in line with these ambitions led them to establish a new system for the dissemination of knowledge, notwithstanding the application of existing security rules. While the knowledge produced in the context of the committee's programs was not intended to be classified, restricted scientific information that would facilitate this production was (to some extent) more openly released. The 'need-to-know' principle occupied an important place in these decisions. A legacy of WWII security, it was now perceived as an obstacle to setting up collaborative projects, since it compartmentalized research procedures and practices, and thereby obstructed international exchanges between scientists. If researchers funded by NATO sought to know more collaboratively, then the knowledge allowing them to do so could no longer be distributed on a strict 'need-to-know' basis.

We have seen that, although decisions on the 'need-to-know' principle and other aspects of information security were taken collegially in the newly established NATO Science Committee, a small number of US science administrators took the lead in addressing the secrecy problem. Ramsey, Robertson and Rabi thought about new procedures and rules that would lessen the negative impact of secrecy and compartmentalization on the committee's proceedings and boost international collaboration.

Their assertion of leadership in discussing the 'secrecy problem' helped these science administrators to inform the production of new NATO-sponsored knowledge, with the additional assistance of precious allies. Steacie propagandized a notion of security through achievement, and the need to go beyond the 'need-to-know' criterion. Rabi and Robertson then seized the opportunity of sponsoring Steacie's report, as it persuaded the other committee's delegates to release more information, but it did not bind any of them to this action. These prescriptions were especially agreeable to the three American officers, because the US scientific-military complex was the largest producer of classified scientific information, and its administration did not wish to share information on atomic weapons with European allies. The offer of information sharing in the study of defense environments helped to circumvent the question of whether to share knowledge on more compelling defense matters. Of course, NATO as a defense alliance continued to harbor contradictions, especially since three of its members (Britain, France and the USA) had undertaken independent nuclear projects. The US administration continued to view the French nuclear program as a major threat to national security, but agreed to share nuclear information with Britain, an element that explains the support Zuckerman provided to Rabi and Ramsey's proposals at the Science Committee meetings.

Zuckerman helped Ramsey to take responsibility for key decisions, such as on the selection of those who decided which defense items ought to be taken into consideration by the committee. Ramsey was thus able to implement new rulings, so that committee members could learn and discuss about the issues which the DRDs considered priorities. In this way, US plans to instigate joint research on issues that would prove to have an impact on air and sea defense could be approved. Now Rabi and the intelligence-savvy Robertson worked towards making available the ASTIA reports, so that the search for more effective means to carry out surveillance operations could instigate the promotion of research on the upper atmosphere (related to the need to track enemy satellites) and oceanic research (in connection with anti-submarine warfare).

The increased distribution and declassification of research materials invigorated scientific cooperation in the study of the defense environment while securing that on nuclear weapons and missiles firmly in US hands. Some of the research projects funded by NATO benefitted enormously from the sharing of ASTIA technical information, since some classified documents were now available to researchers sponsored by the defense alliance. Declassification certainly helped to spread a US model of science-making in geophysical research and led European scientists to embrace it more openly. Furthermore, new security provisions were coupled with the supply of US equipment, which increased standardization and broadened the geographical coverage in carrying out experimental observations.[81]

The management of science and information security at NATO entailed a redefinition of the boundaries of secrecy and openness, in order to accommodate new interests in sharing new knowledge produced in international collaborative endeavors. But the far-reaching implications of this model for Cold War Science have yet to be fully explored. The NATO Science Committee placed the production of open knowledge within a system of circulation of restricted information; this means that the actual uses of the new open knowledge by the defense alliance could be cleverly disguised. It is, therefore, reasonable to assume that, while the new security provisions apparently facilitated the production of open knowledge, they offered ideal cover to projects with unstated defense research goals. Although the NATO science committee never sponsored (in Ronald Doel's terms) 'science in black,'[82] the darker sheen of its 'science in white' will undoubtedly attract historians in the future.

81 For an analysis of this process, see the essays in Turchetti and Roberts (eds.), *The Surveillance Imperative.*

82 Ronald Doel, "Scientists as Policymakers, Advisors and Intelligence Agents," in Thomas Söderqvist (ed.), *The Historiography of Contemporary Science and Technology* (Amsterdam, 1997), pp. 215–44.

CHAPTER 3

A Transnational Approach to US Nuclear Weapons Relationships with Britain and France in the 60s and 70s

John Krige

In July 1958 the American Congress voted in Public Law 85-479. On the face of it, PL 85-479 was an astonishing reversal of prior policy on nuclear secrecy. The severe restrictions imposed by the McMahon Act in 1946 had been relaxed in 1954 to promote Eisenhower's Atoms for Peace program. PL 85-479 went much further. Instead of merely encouraging international cooperation for purely peaceful purposes, it offered American help to enhance the nuclear weapons capability of other nations. Specifically, it authorized the Atomic Energy Commission to cooperate with another country, exchanging with it Restricted Data necessary "to improve its atomic weapon design, development or fabrication capability [...]." This exchange was not unconditional, however. The more liberal approach encouraged by PL 85-479 was only applicable "provided that nation has made substantial progress in the development of atomic weapons."[1]

The PL 85-479 amendment did not emerge in a vacuum: it was President Eisenhower's response to relentless pressure from some of his European allies to gain access to protected scientific and technological information about the US weapons program. Britain had embarked on its A-bomb program in 1947, and on the development of an H-bomb in 1954. France tested its first fission bomb in the Sahara in 1960. The British wanted design and fabrication information on both atomic and hydrogen weapons that would save many years of R&D, and millions of pounds. The French were equally enthusiastic to secure US assistance. Eisenhower, who said that the McMahon Act was a terrible piece of legislation that undermined the US's relationships with its NATO partners, was willing to assist. But he had to contend with Congress and the Congressional Joint Committee on Atomic Energy (JCAE), which was extremely reluctant to share any nuclear secrets. Sharing would, the Committee claimed, encourage proliferation and might leak sensitive information to the Soviet Union via

1 US Congress, Public Law 85-479, July 2, 1958, http://www.gpo.gov/fdsys/pkg/STATUTE-72/pdf/STATUTE-72-Pg276-2.pdf, Section 7.

 | DOI 10.1163/9789004264229_004

espionage: the UK was seen as being notoriously soft on security, while the French CEA was supposedly riddled with communists. By 1958 Eisenhower had managed to overcome Congressional resistance, and PL 85-479 was passed. That law deliberately favored the UK over other international partners since it was deemed to have "made substantial progress in the development of atomic weapons." France would wait 15 more years before a new Republican president and his national security adviser, to wit Richard Nixon and Henry Kissinger, no longer had these concerns and extended many of the same privileges to France that had been given to Britain—carefully concealing this fact from the JCAE and the British.

There is an increasing emphasis in contemporary scholarship on the Cold War as a global phenomenon that burst the bounds of the national container, that transcended rivalry between superpowers, and that actively engaged European allies (albeit on American terms), at least in the 1950s and 1960s. This paper shares this perception, and argues that it applies, with qualifications, even to the transnational circulation of knowledge on nuclear weapons. Taking its cue from Jim Secord's claim that the how and the why of knowledge circulation are the central questions of our field, this paper describes the 'how' and 'why' of the circulation of knowledge about nuclear weapons between the US and Britain and France at different phases of the Cold War.[2] It is noteworthy that, while historians of science and technology have shown considerable interest in the regulatory regime surrounding nuclear knowledge, their approach has seldom broken out of the national frame. This paper situates America in the world at the hub of a network of nuclear weapons knowledge circulation, and explores the circumstances under which top American officials encouraged cooperation and exchange in this most sensitive of all domains.

Privileging the UK

In 1955 a small cadre of experts chaired by a senior civil servant, William Strath, described the effects of a carefully targeted Soviet attack on the United Kingdom. Their secret document evaluated the damage inflicted by just ten 10kt hydrogen bombs that made optimum use of prevailing winds to spread radioactive fallout far and wide. As Strath explained to the government, the result would be dire:

2 James A. Secord, "Knowledge in Transit," *Isis* 95 (2004), 654–72.

> half the population killed, major cities laid waste and the rest of the country and its people subjected to enough radioactive fallout to contaminate huge areas and kill millions more. Agriculture and communications would be devastated, and industry and the economy destroyed. Fire fighting, medical, transport, water and food supply services, and even the machinery of government itself, would likely collapse, and with them civil society. The immobilized and immiserated survivors would struggle alone against disease, starvation and the psychological effects of nuclear bombardment. Britain could not survive as a platform in a thermonuclear war, and its very survival as a nation would be in doubt.[3]

The British promptly embarked on the construction of six new nuclear reactors to fuel an H-bomb program. They hoped that, once their intentions were known and their results looked promising, they would be able to count on the US for collaborative support. This would reverse years of exclusion from the American nuclear effort that had begun with the passage of the McMahon Act. The logic was similar to that which had persuaded the British to embark on their own atomic bomb program in 1947: a successful independent effort would confirm Britain's status among the great powers. And with something of technological interest to offer, they might get something in return.[4]

The political obstacles to this plan were formidable. Eisenhower saw the McMahon Act as an impediment to allied collaboration and burden sharing. Nevertheless, time and again his hopes of engaging Britain were thwarted, whether the issue was sharing data on submarine nuclear propulsion, or on the weight and dimensions of American weapons carried by British bombers, or control over the launch of sixty nuclear tipped Thor missiles that were installed in silos around the country in the late 1950s. The administration was hesitant to share delicate nuclear secrets with a country that had been shaken by a series of spy scandals, beginning with the exposure of Klaus Fuchs' treachery in 1950.

A number of factors turned the tide in late 1950s. The launch of Sputnik in October 1957 exposed the US's vulnerability to an attack from beyond its shores. The President's refusal to support the disastrous Anglo-French-Israeli invasion of Egypt in November 1956, after President Nasser nationalized the

---

3 Jeff Hughes, "The Strath Report: Britain Confronts the H-bomb, 1954–1955," *History and Technology* 19 (2003), 257–75.

4 Timothy J. Botti, *The Long Wait. The Forging of the Anglo-American Nuclear Alliance* (New York, 1987); Jan Melissen, *The Struggle for Nuclear Partnership: Britain, the United States and the Making of an Ambiguous Alliance* (Groningen, 1993).

Suez canal, called for bridge-building with a faithful ally. Finally, a series of successful UK tests of hydrogen bombs from 1957 onwards led weapons scientists to believe that there would not be a one-sided transfer of knowledge from America to Britain after all. American eyewitnesses at the Grapple series in 1958 believed that the UK had important and useful knowledge to share on the development of thermonuclear weapons.

It is often said that the intimacy of US–UK nuclear collaboration derived from a 'special (emotional) relationship' based on a shared history, culture and language. John Baylis insists that more substantial elements also served to bond the two countries together, in particular the reciprocal exchange of advanced knowledge about the design of thermonuclear weapons.[5] Soon after US scientists witnessed the Grapple tests, expert groups from both sides met at the Sandia and Lawrence Livermore labs in the US, and at Aldermaston in England. One report of these meetings claimed that "[i]nformation streamed across the Atlantic at a breathtaking pace."[6] The British saved much time and money, and undoubtedly improved the technical aspects of their nuclear strike force, thanks to American information on items like weapons electronics; uranium and plutonium fabrication; on the properties of tritium and beryllium; and on the assembly of weapons. The US is not likely to have gained much detailed knowledge from the British, but the exchange still proved educational for them. Those American scientists and engineers who discussed weapons design with their British peers drew lessons from the paths their colleagues abroad had taken to solve nuclear weapons problems. Thus the US benefitted by virtue of what I shall call 'positive guidance' from the British, rather than from the transfer of fundamentally important new information. This at least was the justification for the new policy that the deputy assistant for European affairs, Frederick Jandrey, made to the JCAE in July 1958:

> For us, the merging of British information and know-how in a common fund with our own [...] should provide two direct benefits. Since nuclear weapons programs have been carried out independently by each country we stand to gain from techniques developed by the British where they have solved the same problems which we faced by methods different from our own. For the same reason, where we find that British techniques developed separately are closely similar to those we have evolved, we can

5 John Baylis, "Exchanging Nuclear Secrets: Laying the Foundations of the Anglo-American Nuclear Relationship," *Diplomatic History* 25 (2001), 33–61.

6 Baylis, "Exchanging Nuclear Secrets," p. 53.

> have added confidence in the evolution of our own program. Thus our weapons program should profit measurably through the stimulation which inevitably results from the cross-fertilization of ideas.[7]

It is evident, and understandable, that the UK gained far more from this exchange than did the US. Baylis's emphasis on 'reciprocity' is exaggerated if understood as an exchange of equivalent types and amounts of information. But then his aim was to counter the view that cultural affinity animated the special nuclear relationship. What he does not emphasize enough, though, is that the US shared sensitive knowledge not simply in order to learn what (little) they could from the British about nuclear weapons design: they also used the obvious discrepancy in power (in their favor), in combination with Britain's pressing need for nuclear knowledge, to lever separate, different political ends. For example, Washington sought assurances that London would not again undermine the progress towards European integration enshrined in the Treaty of Rome that came into force in January 1958. It wanted the UK to follow its line in the UN and agree that Mao's China should not be represented in the general assembly. This *linkage* between sharing nuclear knowledge and quite distinct non-nuclear goals is an expression of the soft power that was available to the US by virtue of its nuclear weapons leadership—a leadership that provided it with a political instrument to shape the policies of its partners in line with its interests.

US—French Nuclear Relationships

In 1989 Richard Ullman, a professor of international affairs at Princeton University, revealed that the US had provided substantial covert, and likely illegal, nuclear assistance to France for 15 years. His claims were based on over 100 interviews with top-level politicians, civil servants, scientists, industrial managers and military and naval officers in both countries. He noted that, although the exchanges avoided discussing the actual design of nuclear weapons—unlike in the case of the British—the French did acquire a large amount of highly sensitive information.

The publication of Ullman's paper was met with shock and incredulity. It was hard to believe that Nixon and Kissinger would help France build an independent nuclear deterrent, upsetting years of established policy. In October

7 Baylis, "Exchanging Nuclear Secrets," p. 59.

1957, with France set on the nuclear path, Macmillan and Eisenhower agreed that they would not actively hinder the French nuclear program, but that they would not help it either.[8] The climate changed under Kennedy and Johnson. US policy hardened after the successful French nuclear test in February 1960 and continued into the decade. President de Gaulle's increasing determination to acquire an independent deterrent as a symbol of French sovereignty, his dislike of the European Community and of NATO, and his withdrawal of French troops from under NATO command along with his demand that all US forces leave French soil, did not endear him to Washington. In 1964 McGeorge Bundy signed off on NSAM 294 (National Security Action Memorandum). It stated that, given France's reluctance to integrate its independent nuclear deterrent into NATO,

> [i]t continues to be in this government's interest not to contribute to or assist in the development of a French nuclear warhead capability or a French national strategic nuclear delivery capacity. This includes exchanges of information and technology between the governments, sales of equipment, joint research and development projects, and exchanges between industrial and commercial organizations [...].[9]

This policy remained in place after de Gaulle left his position of power in April 1969 to be replaced by Georges Pompidou. Pompidou, while more pragmatic than the General, still strongly believed in French autonomy. How, then, could Ullman's claims be true? The bubble burst when the next President, Valery Giscard d'Estaing, who took office in 1974, published his memoirs in the early 1990s and acknowledged that such exchanges had taken place. The authenticity of Ullman's story has been confirmed by the subsequent release of formerly classified documents, most notably a batch of about 50 items that were recently published online by William Burr of the National Security Archive.[10] We now know that there were, indeed, extensive exchanges between American

---

8 See also Maurice Vaïsse, "Un dialogue de sourds. Les relations nucléaires franco-américaines de 1957 à 1960," *Relations internationals* 68 (1991), 407–23.

9 National Security Action Memoranda 294, 20 April 1964, "US Nuclear and Strategic Missile Delivery Assistance to France," available at http://www.lbjlib.utexas.edu/johnson/archives.hom/nsams/nsam294.asp [accessed 16 May 2014].

10 William Burr, "US Secret Assistance to the French Nuclear Program, 1969–1975: From 'Fourth Country' to Strategic Partner," available at http://www.wilsoncenter.org/publication/us-secret-assistance-to-the-french-nuclear-program-1969-1975-fourth-country-to-strategic [accessed 16 May 2014].

and French high-level officials and scientists and engineers about nuclear and missile technology between 1971 and 1975—and that these exchanges are likely to have continued beyond that period, though that material remains classified.

Ullman's paper caused such a sensation in part because the actors had managed to keep the secret so well. In fact, secrecy was such a major preoccupation that no high level meeting was held without ensuring that its existence could be covered up. Here, for example, is an extract from the minutes of a meeting held on the patio at the Western White House in San Clemente on 31 August 1973. Henry Kissinger, Nixon's National Security Adviser, was host to the French Minister of the Armed Forces, Robert Galley. The official minutes of their encounter start as follows:

> Kissinger: If anything should get out to the press, we'll just say that you stopped here on your way back from Tahiti.
>
> Galley: We took every precaution to ensure secrecy. I am sure no one knows on our side.
>
> Kissinger: The trouble on our side is that the Air Force people see your plane.
>
> Walters: We can just say the plane landed here because the other field was closed. It was to land at March but it had to go to El Toro instead.
>
> Kissinger: We'll say you are an old acquaintance, your plane was diverted to El Toro, you called to pay your respects, and I asked you down. There is very little possibility that this will happen, but I like to be prepared.
>
> Galley: The press is not interested in things over weekends.[11]

These elaborate procedures are indicative of the sea change in policy that was under way.[12] The French required information and technology to strengthen the core of their independent deterrent. Kissinger was willing to lend a helping hand because, now that the *force de frappe* was a reality, he believed that the US might as well integrate it into its strategic planning and have France participate seriously in burden sharing. However, this required that existing technologies be upgraded. In their then-current state they were not just of little use, but positively dangerous: they were so vulnerable that they could only be

11 Memorandum of Conversation at San Clemente, 31 August 1973, in Burr, "Secret Assistance," Document 46.

12 Pierre Melandri, "Aux origines de la coopération nucléaire franco-américaine," in Maurice Vaïsse (ed.), *La France et l'atome: Etudes d'histoire nucléaire* (Brussels, 1994).

effective if used pre-emptively—and no one wanted France to trigger a nuclear war with the Soviet Union because it was equipped with obsolete material. The problem was that neither Congress nor the JCAE shared Kissinger's vision for nuclear cooperation with France. For this reason the discussions about sharing sensitive knowledge with Paris, and the concrete activities in which such knowledge was actually exchanged, were shrouded in secrecy, and called forth the most elaborate gymnastics to avoid them being brought down by hostile domestic forces.[13]

The flow of knowledge from Washington to Paris was irregular, and came in bursts. This was partly because of US prudence, particularly as French demands for more, and more sensitive information escalated. It was also affected by fluctuations in the warmth of Franco-American relations.[14] 1973 was the nadir. Kissinger announced a 'Year of Europe' that was intended to consolidate transatlantic bonds; however, it had the opposite effect, due to the accumulated repercussions of a number of major decisions by the Nixon administration that were taken without consulting their allies (suspension of the convertibility of the dollar, the SALT agreements), and because of serious differences over how to handle the six-day war in the Middle East. Kissinger was incensed: the 'Year of Europe' initiative, "already widely and wildly misinterpreted in Europe as blackmail, hegemonism and an effort to force Europe into a regional role," had offered far more to Europe than any other administration "would or could" do.[15] Kissinger developed a visceral contempt for one senior French Minister, Michel Jobert, who made particularly hostile comments about US behavior. At a meeting of the North Atlantic Council in March 1974 he went so far as to state publicly: "In this room there is a representative of a country who is doing all he can to obstruct American policy. I want him to know that, if his country persists in this behavior, I'll crush it."[16]

---

13 Kissinger's 'backchannel diplomacy' is critically analyzed by Tony Judt in his "The Illusionist: Henry Kissinger and American Foreign Policy," in *Reappraisals. Reflections on the Forgotten Twentieth Century* (New York, 2008), chapter 20. I would like to thank Jessica Wang for providing this reference for me.

14 Georges-Henri Soutou, "Le Président Pompidou et les relations entre les États-Unis et l'Europe," *Journal of European Integration History* 6 (2000), 111–46; Marc Trachtenberg, "The French Factor in U.S. Foreign Policy during the Nixon—Pompidou Period," *Journal of Cold War Studies* 13 (2011), 4–50.

15 Helmut Sonnenfeldt to Henry Kissinger, Memorandum "Supplementary Checklist for Meeting with French Defense Minister," Top Secret, 26 July 1973, available in Burr, "Secret Assistance," Document 40.

16 Quoted in Maurice Vaïsse, *La Puissance ou l'influence? La France dans le monde depuis 1958* (Paris, 2009), p. 194.

Two National Security Decision Memoranda (NSDM 103, 104), which were authorized by Kissinger in March 1971, defined what help the US could give.[17] They covered the export of high-powered computers, and assistance with both land- and submarine-based ballistic missiles, and with nuclear safety. These were subsequently extended to include assistance with nuclear weapons per se. Each of these merits a closer look.

The export of advanced computers for use in the French weapons laboratories was deftly dealt with: they were simply recategorized. As NSDM 103 put it, while "we cannot rescind entirely our restrictions on the export of advanced computers for use in French nuclear weapons laboratories [...], we will redefine 'advanced computers' so that in practice some of the models currently falling under the restrictions on end-use will become available without any restrictions." This new definition allowed the unimpeded export of all computers whose power was rated below that of the IBM360/165. More powerful computers could be exported, but required a special letter assuring the State Department that they would not be misused. In practice, this allowed the French to buy four CDC7600 computers by 1975. The brainchild of Seymour Cray, these were the most powerful devices on the market in the early 1970s. Such sales were conditional on the machines not being used by the weapons labs, though everybody knew that they would be. In 1975 the French made a formal request to import a CDC7600 for military use since, as one senior French official put it, they "would much prefer to have one that they could use with a clear conscience."[18]

NSDM 103 offered French missile assistance on the condition that the information shared would not jeopardize the security of the US's own weapons programs, or provide France with what was called a 'distinct new capability' in areas like guidance, missile accuracy and re-entry vehicle hardening. By August 1973 there had been exchanges of information, among other things, on propulsion; on stabilizing submarine gyros that tended to drift; and on securing missiles from an accidental launch from a submarine. Overall, French anxieties were reduced with regard to reliability, safety, deterioration, vulnerability,

17 National Security Decision Memorandum (hereafter NSDM) 103, 29 March 1971, "Military Cooperation with France," available at https://www.fas.org/irp/offdocs/nsdm-nixon/nsdm-103.pdf; NSDM 104, 29 March 1971, "Cooperation With France on Nuclear Safety," available at https://www.fas.org/irp/offdocs/nsdm-nixon/nsdm-104.pdf [accessed 16 May 2014].

18 Helmut Sonnenfeldt, Memorandum for the Record, "Conversation with Delpech," Top Secret, 9 October 1975, in Burr, "Secret Assistance," Document 55.

and hardening to withstand the effects of other nuclear blasts.[19] The US was willing to assist with MRVs (Multiple Re-entry Vehicles, or many warheads on one missile) but not with MIRVs (the independently targeted version), since the latter had been at the core of US-USSR negotiations over the SALT I agreements. In 1975 President Gerald Ford extended US technical assistance for the new generation of French submarine-launched missiles. This included basic research into hardening the missile, for example by exposing its materials to underground nuclear tests in Nevada, and to plasma jets, in order to simulate the erosive conditions of re-entry.[20]

Assistance with nuclear weapons design (the final example this article will examine in more detail) was, obviously, the most sensitive topic of all. PL 85-479 could, in principle, be used to authorize assistance to the French in this domain, since it could be argued that the country had made considerable independent progress with its own nuclear weapons. However, this would require formal permission from the JCAE and, once made public, would doubtless cause an uproar in Congress. This is something that Kissinger wanted to avoid at all costs. But he still wanted to be cooperative. He came up with the solution to this dilemma when told that the French were having trouble deciding on the best design for a booster trigger in their thermonuclear weapons. Kissinger replied:

> We may not be able to give you information, but we can critique what you are doing. We can say "That's the wrong way." So there are many ways to give you information. [...] It can be like a seminar; you can say you have three possibilities and we can tell you, "that's wrong; that's complicated," etc.[21]

This procedure, now famously known as 'negative guidance,' was labeled 'Twenty Questions' by the American scientists and engineers who were interviewed by Ullman.[22] Simply by letting the French know whether or not they were on the right track, the US experts could both assist them and not reveal restricted data. Kissinger invented this procedure to ensure that, if needs be,

---

19 Helmut Sonnenfeldt to Henry Kissinger, Memorandum, "Supplementary Checklist for Meeting with French Defense Minister," Top Secret, 26 July 1973, in Burr, "Secret Assistance," Document 40.

20 Sonnenfeldt, "Conversation with Delpech."

21 Memorandum of Conversation at San Clemente, 31 August 1973, in Burr, "Secret Assistance," Document 46.

22 Richard Ullman, "The Covert French Connection," *Foreign Policy* 75 (1989), 3–33, p. 10.

he had a "domestic situation we can defend:" critiquing the French program in this way did not, in his opinion, require Congressional approval.

Why did Kissinger go to such lengths to assist an ally that was often anything but grateful for the much-required help? One reason, which was touched upon above, was that, now that a French nuclear force existed, "it was important that it does not become irrelevant." France needed to be "strong from a military, moral and political point of view. We think that if a major French military program collapses it would be very bad, because then the French alternative would be neutralism," as Kissinger explained to Galley.[23]

But there was more to it than that: Kissinger's determination to help France was part of a more general shift in attitude in the White House. In the early 1970s both he, and to some extent Richard Nixon, recognized that their country's power was not unlimited—the debacle in Viet Nam and the oil crisis made that plain. In this context, to quote Kissinger's most recent biographer Jeremi Suri, they made it their goal to "transform the structure of the international system to encourage a diffusion of power on terms favorable to the United States." A world with more "centers of decision," i.e. a more multipolar world, Kissinger believed, would provide flexibility for innovative diplomacy and consensus building. "In Western Europe," in particular, Suri points out, this more federal approach to American leadership "would allow Washington to draw more effectively on the potential strength of its allies, and would ensure their firmer and more self-confident engagement with the White House."[24]

The emergence of a potentially strong bloc in Western Europe was antithetical to an agenda that was premised on a diffusion of power among nation states in the region. On 1 January 1973 the European Community had been enlarged by the inclusion of the United Kingdom, along with Denmark and Ireland. The American authorities were disturbed by this step towards the consolidation of what might eventually become a third way between capitalism and communism. Kissinger explained to his advisers at the time that he was devising 'negative guidance:' "We must break up the Europeans. And the French are essential." And "we are going to try to bust the Europeans. The French can be useful in this." Giving negative guidance was one of a number of strategies devised to give the French something that "looks like a step

23 Memorandum of Conversation, Top Secret, 27 July 1973, in Burr, "Secret Assistance," Document 41, pp. 6–7.

24 Jeremi Suri, "Henry Kissinger and the Geopolitics of Globalization," in Niall Ferguson, Charles S. Maier, Erez Manela and Daniel J. Sargent (eds.), *The Shock of the Global: The 1970s in Perspective* (Cambridge, MA, 2010), 173–88.

forward but doesn't give them anything yet."[25] Put even more bluntly: "What we want is something which makes Galley drool but doesn't give him anything but something to study for a while. I will brutalize Galley." Kissinger also asked his Defense Secretary Schlesinger and his Deputy Assistant Brent Snowcroft to: "Lead them on without giving up anything—we want to get a handle on them without [their] knowing it."[26] In short, in spite of his intense anger at the behavior of some senior French officials, Kissinger was determined to play to France's desire for independence, both to weaken the power of the growing European bloc that now included Britain, and to integrate the French nuclear deterrent more closely into that of the Atlantic community. As he put it to Schlesinger, if the US could develop a productive bilateral relationship with Galley, the other Europeans would turn on them and say, "you bastards, you talk about unity, and then you go this bilateral route" with the United States.[27]

Were there *quid pro quos* involved in sharing knowledge? Of course. At the most general level, as Kissinger phrased it, "the real quid pro quo is the basic orientation of French policy."[28] Other conditions were also put on the table. One particularly sore point, that seems to have been conceded by the French, was the demand that they settle American claims for the costs of relocating US forces in 1966, when de Gaulle made them leave France.[29] At the same time it was understood that *quid pro quos* for profiting from US advanced knowledge should not be too explicit, nor should "we give [the French] any reason to feel they can get something without continuing to give something in return."[30] That 'something' should also be defense-related. There was a distinct desire not to link increased military assistance to contentious, non-defense policy issues, e.g., disagreements over tariffs and trade.

While the French were immensely grateful for all the help they received, and said as much many times at an official level, they were also very careful

25 Memorandum of Conversation, "Visit of French Defense Minister Galley: Strategic Programs," Top Secret, 17 August 1973, in Burr, "Secret Assistance," Document 43.

26 Memorandum of Conversation, "French Nuclear Discussion," Top Secret, 9 August 1973, in Burr, "Secret Assistance," Document 42.

27 This quote is from Trachtenberg, "The French Factor," p.39. Further, "We want to keep Europe from developing their unity against us. If we keep the French hoping they can get ahead of the British, this would accomplish our objective," ibid.

28 Memorandum of Conversation, Kissinger and Schlesinger, Top Secret, 5 September 1973, in Burr, "Secret Assistance," Document 47.

29 Memorandum from Helmut Sonnenfeldt to Henry Kissinger, "Missile Assistance to France—New NSSM," 3 February 1973, in Burr, "Secret Assistance," Document 37.

30 Sonnenfeldt to Kissinger, "Missile Assistance to France," loc. cit.

about the information they revealed about their program. They were suspicious of American intentions. The kinds of concerns they had were explained by one or two senior officials at an eye-witness seminar in Paris in 1994 and in other memoires. They rejected US offers to test their weapons underground in Nevada: they did not want to disclose details about their weapons design and performance. They suspected that the logjam over the supply of the CDC7600 had been broken because the US hoped thereby to gain access to the movements of their nuclear submarines. One very senior official even claimed that US scientists and engineers had deliberately fed them false information during one of the Twenty Question sessions, to send them down the wrong track. Pierre Messmer, Defense Minister under de Gaulle and Prime Minister for about two years beginning in May 1972, offered the most negative perspective: "I was never disappointed since I had no illusions as to what we could expect from nuclear weapons relations with the Americans [...]. Certainly the US wanted to keep up the contacts, and they judged that the best way to do that was to imply that they could give us something useful. In fact, we quickly realized that they would either give us nothing at all, or only what we had already."[31] There is clearly some truth in this statement—after all, Galley was supposed to drool but not to be given too much; other evidence suggests that this judgment is somewhat exaggerated. In any event, the US–French nuclear relationship was far more rocky than the special Anglo-American relationship. Suspicion and mistrust on both sides surely limited the degree of cooperation.

Concluding Remarks

The themes discussed in this essay can be connected to broader questions in our field. It extends my two recent studies on US-UK relationships in the domain of classified gas centrifuge enrichment technology.[32] It also takes a transnational approach to the circulation of sensitive knowledge between the

31 Maurice Vaïsse, "Les 'relations speciales' franco-américaines au temps de Richard Nixon et Georges Pompidou," *Relations Internationales* 119 (2004), 345–62.

32 John Krige, "Hybrid Knowledge. The Transnational Co-production of the Gas Centrifuge for Uranium Enrichment in the 1960s," *British Journal for the History of Science* 45 (2012), 337–57; John Krige, "U.S. Technological Superiority and the Special Relationship. Contrasting British and American Policies for Controlling the Proliferation of Gas Centrifuge Enrichment," *The International History Review* 36 (2014), 230–51.

US and its European allies, with the specific intention of exploring both the micro-structures of knowledge exchange and the power relations in which they are embedded, relations that regulate and set limits on just what can be shared. It thus resonates with Jim Secord's insistence that the 'how' and the 'why' of the circulation of knowledge is one of the central questions of the history of science; it also responds to Hunter Heyck and David Kaiser's recent call for a global approach to the history of science in the Cold War, and one that transcends the binary logic of superpower rivalry.[33]

The analysis combines the history of science with nuclear history and the history of foreign policy. It is worth emphasizing that this understanding of foreign policy, while posing new challenges to historians of science, is essential for meaningful transnational or global studies of the flow of nuclear science and technology. It is not simply that it provides 'context'. An understanding of foreign policy is methodologically essential, since it injects the state as a regulator of knowledge flows into the core of the global system. As such, it fulfills a dual purpose: it rejects a highly ideological concept of the global that effaces the state altogether and speaks of a world without boundaries; and it solves the political (if not epistemological) problem of how to articulate the local with the global: the state as regulatory mechanism defines the physiognomy of the global, and so the terms on which collaborative networks can be established.

Steering the argument to describe what I conveniently labeled 'positive guidance' in the British case, and 'negative guidance' for the French, expresses a deliberate methodological choice. It highlighted some of the micropractices whereby knowledge was exchanged in face-to-face encounters between American scientists and engineers, and their allies. As such it is a gesture to Stephen Shapin's emphasis on the importance of mundane processes in the circulation of knowledge.[34] On a more fundamental level, it provides a hook into Kapil Raj's analysis of cross-cultural circulation between British travelers and Indian sages.[35] Raj insists that circulation itself is a site of knowledge formation. He thus refuses to see circulation in terms of a linear model that speaks of production followed by dissemination, followed in turn by imposition, selective appropriation or rejection. His concept of circulation is transformative, i.e. it focuses on the process of knowledge-making in the encounter,

33 Hunter Heyck and David Kaiser, "Introduction: New Perspectives on Science and the Cold War," *Isis* 101 (2010), 362–6.

34 Stephen Shapin, "Rarely Pure and Never Simple: Talking About Truth," *Configurations* 7 (1999), 1–14.

35 Kapil Raj, *Relocating Modern Science* (London, 2008).

bestows agency on both parties in that engagement, and teases out the processes of power and resistance, negotiation and reconfiguration, that occur in face-to-face interactions. This approach has led me to draw attention to positive and negative guidance as strategies for knowledge exchange. It is central to my emphasis on the reciprocity that makes such exchanges possible, and it has led me to avoid using mechanical terms like diffusion. It also made me focus on the mutual distrust that prevailed between the US and their partners as knowledge circulated transnationally between them.

This paper has further expanded the locus of power in America's relationships with its partners, the mechanism whereby hegemony is coproduced, to use some of my earlier language: it has enlarged the site of power to include both the differential in knowledge itself and the linkages that provide the *quid pro quos* for its circulation. Power is expressed in implementing these links, that instrumentalize knowledge as a political weapon to shape the orientation of British and French nuclear policy. Correlatively Kissinger's determination not to 'overplay the card,' the discussion with his advisers on the kind and reach of the linkages, and the limitation on the *quid pro quos* for providing negative guidance are indicative of his perception of the US as living in an interdependent world—a world in which its power was increasingly limited. It is a marker of the relative decline of American power and hegemony that began in the 1970s with the onset of globalization.

PART 2

Dutch Perspectives

CHAPTER 4

Putting a Lid on the Gas Centrifuge: Classification of the Dutch Ultracentrifuge Project, 1960–1961

Abel Streefland

In recent years a growing interest has developed in the history of classified and secret science. Classification of knowledge has been shown to be a phenomenon of all times and all places, ranging from early modern navigational maps which were regarded as state secrets in Portugal and Spain, to the secret recipe for Coca Cola syrup.[1] Secrecy was particularly pertinent in the context of Cold War nuclear research, much of which was classified as secret.[2]

The classification of Cold War science not only affected the production and circulation of knowledge in that period, it also poses particular difficulties to the present-day historian aiming to reconstruct its history. Large parts of the relevant archives are still classified. Such prolonged secrecy also applies to the Dutch ultracentrifuge project, the classification of which will be the focus of this article.[3] The archive of the main actor, Jaap Kistemaker, was moved, following his retirement in 1982, from his laboratory in Amsterdam to Almelo, where it was secured within the walls of Urenco, a company that produces enriched uranium. Kistemaker's archive is difficult to access, as it contains technical specifications that are still judged to be highly sensitive. Access, if

1 For an introduction to the subject see Koen Vermeir and Dániel Margócsy, "States of Secrecy: An Introduction," *The British Journal for the History of Science* 45 (2012), 153–64.

2 See for example Peter Galison, "Removing Knowledge," *Critical Inquiry* 31 (2004), 229–43; Peter Galison, "Secrecy in Three Acts," *Social Research* 77 (2010), 941–74; Alex Wellerstein, *Knowledge and the Bomb: Nuclear Secrecy in the United States, 1939–2008* (PhD dissertation, Harvard University, 2010).

3 So far little has been published on the Dutch ultracentrifuge program. Some literature touches upon the matter: Susanna Schrafstetter and Stephen Twigge, "Spinning into Europe: Britain, West Germany and the Netherlands 1970," *Contemporary European History* 11 (2002), 253–72; R. Scott Kemp "The End of Manhattan: How the Gas Centrifuge Changed the Quest for Nuclear Weapons," *Technology and Culture* 53 (2012), 272–305; William Burr, "The 'Labors of Atlas, Sisyphus, or Hercules'? US Gas-Centrifuge Policy and Diplomacy, 1954–60," *The International History Review*, article online since 23 May 2014 (27 pp.); the forthcoming PhD dissertation by Elmar Hellendoorn (Utrecht University) will discuss this episode from the perspective of the Dutch Ministry of Foreign Affairs.

 | DOI 10.1163/9789004264229_005

permitted at all, requires security clearance. Even if this hurdle is taken, permission to publish from these archives may still be withheld.

At first sight 'secret science' may seem an oxymoron. Science is usually associated with ideals of openness, transparency and communalism, all aspects supposed to be essential to its functioning properly.[4] Yet, when science is classified, it is fenced off from (parts of) the academic community. Publication of results becomes problematic, and exchanges with other specialists outside of the local research environment may become impossible. In this sense, classified science does not work in the same way as 'normal' science does. Different rules apply to the production and circulation of classified knowledge. Because of the supposed antagonism between science and secrecy, classification of science is usually portrayed as an external condition, imposed upon reluctant scientists or engineers by either governments or industries.[5]

In the case at hand, it was the Dutch government that decided to classify Dutch research on the enrichment of uranium by means of ultracentrifuges as secret in 1961. They thereby followed a request from the US government, which was based on the risk for proliferation of sensitive centrifuge knowledge. At first sight, this state of affairs fully conforms to the familiar picture of the relationship between the US and Western Europe during the early Cold War period. Generally, the US aimed to stay in control of technological developments in Europe. To this end they used several strategies, most of which can be subsumed under the label 'soft power.'[6] Requesting classification was one of these strategies, as was recently argued by John Krige in relation to the development of gas centrifuges by American and British scientists and engineers.[7]

4 The above reflects, of course, Robert Merton's 'CUDOS.'

5 Galison, "Secrecy in Three Acts."

6 John Krige, *American Hegemony and the Postwar Reconstruction of Science in Europe* (Cambridge, MA, 2006); John Krige, "Building the Arsenal of Knowledge," *Centaurus* 52 (2010), 280–96; John Krige, "The Peaceful Atom as Political Weapon: Euratom and American Foreign Policy in the Late 1950s," *Historical Studies in the Natural Sciences* 38 (2008), 5–44; John Krige, "Atoms for Peace, Scientific Internationalism, and Scientific Intelligence," *Osiris* 21 (2006), 161–81; Jaap van Splunter, *Kernsplijting en diplomatie: de Nederlandse politiek ten aanzien van de vreedzame toepassing van kernenergie, 1939–1957* (Amsterdam, 1993); see also David Baneke's article in this volume.

7 Recently a lot of work has been done on British and American policies with regard to ultracentrifuges. This work mainly focusses on the late 1960s, when cracks started to appear in the special relationship between the two countries; see John Krige, "US Technological Superiority and the Special Nuclear Relationship: Contrasting British and US Policies for Controlling the Proliferation of Gas-Centrifuge Enrichment," *The International History Review* 36 (2014), 230–51; John Krige, "Hybrid Knowledge: The Transnational Co-Production of the Gas Centrifuge

The fact that the Dutch fully complied with the American request, apparently without strong pressure, confirms the picture of 'consensually co-constructed hegemony.'

However, a more subtle image emerges when we start to focus on the reactions of different parties to the American request of 1960 to classify Dutch gas centrifuge technology. In this article, I will show that, in this specific case, the classification of knowledge did not follow the straightforward path suggested by the outcome. The American request induced lengthy debates involving representatives of different powers: politics, science, technology and industry. These all had their own interests, and accordingly, their own reservations against the American proposal. Economic and political interests vied with scientific and industrial concerns. It was generally feared that secrecy would harm the project, impede cooperation between science and industry, and obstruct the intended collaboration with the German centrifuge group. Furthermore, the division of roles between the actors was somewhat unusual. On the one hand, Dutch politicians showed reluctance to press matters, especially as they came to distrust the American motives for secrecy. They wondered whether proliferation was really the Americans' main motivation, or if economic and industrial interests dictated their agenda. On the other hand, the scientists were less averse to the proposed regime of secrecy than one might expect. In fact, they had found a way to turn the act of classification to their advantage. In this somewhat uncertain situation, contingent events, particularly the German decision to classify centrifuge projects, may have been the deciding factor in the Dutch parties' final decision to grant the American request.

The Dutch Ultracentrifuge Project

Research on uranium enrichment in the Netherlands started immediately after World War II. The driving force behind the endeavor was Jaap Kistemaker, a physicist working for the newly founded Foundation for Fundamental

for Uranium Enrichment in the 1960s," *The British Journal for the History of Science* 45 (2012), 337–57; John Krige, "Maintaining America's Competitive Technological Advantage: Cold War Leadership and the Transnational Coproduction of Knowledge," *Humana.Mente* 16 (2011), 33–51; Stephen Twigge, "A Baffling Experience: Technology Transfer, Anglo-American Nuclear Relations, and the Development of the Gas Centrifuge 1964–70," *History and Technology* 19 (2003), 151–63.

Research on Matter (FOM).[8] In 1952 he managed to separate uranium isotopes by using a large electromagnetic isotope separator. Three years later he switched to ultracentrifuges: fast spinning cylinders in which the lighter uranium isotope can be separated from the heavier one. During the war these devices had been discarded by the US in favor of gaseous diffusion plants. This situation changed in the first half of the 1950s. New, stronger alloys became available and techniques for fast-spinning cylinders improved. In 1955, after attending a talk by Hermann Gerhard Hertz in Hamburg on ultracentrifuges that were able to separate isotopes, Kistemaker realized that gas centrifuges might become economically feasible after all. His plan was to assess this technique to see if it was possible to develop a uranium-enriching industry.[9]

From its establishment in 1955 the project was coordinated by Kistemaker, under the auspices of both the board of directors of FOM and the Dutch Reactor Centre (RCN). Funding was provided by RCN whose main goal was to develop a Dutch atomic energy reactor.[10] As uranium enrichment was a small part of the complete reactor cycle, the centrifuge project was only one of many projects sponsored by RCN. In the course of the following years, money for the project was provided on an annual basis, and a decision on its long-term future was not taken until well into the 1960s. This uncertainty naturally led to tensions between RCN and Kistemaker, who thought of his project as much more valuable than the funding body's incremental support seemed to suggest. Nonetheless, Kistemaker always received sufficient funds to continue the project.[11]

Kistemaker's research took place at two locations in Amsterdam. The first was his own laboratory, the so-called 'FOM laboratory for mass spectrometry.'[12] Theoretical research on gas centrifuges was done in this lab. Founded as the first of a number of FOM institutes, its official purpose was to build a mass spectrometer. Enriching uranium was, nonetheless, the primary research

8 For the early history of FOM, see: Friso Hoeneveld and Jeroen van Dongen, "Out of a Clear Blue Sky? FOM, The Bomb, and The Boost in Dutch Physics Funding after World War II," *Centaurus* 55 (2013), 264–93.

9 Jaap Kistemaker, "De ultra centrifuge en zijn mogelijkheden bij het produceren van verrijkt uranium op industriële schaal," 1955, J. Kistemaker Ultracentrifuge Archive, Almelo, Enrichment Technology (private company; hereafter ET), box 6.

10 For the history of RCN, see: Jaap A. Goedkoop, *Een kernreactor bouwen: geschiedenis van de Stichting Energieonderzoek Centrum Nederland. Deel 1: periode 1945–1962* (Bergen, 1995); Cornelis D. Andriesse, *De republiek der kerngeleerden* (Bergen, 2000).

11 Jaap Kistemaker, *De geschiedenis van het Nederlandse ultracentrifuge project* (Amsterdam, 1991); ET, boxes 3 and 11.

12 This laboratory has changed its name a number of times; today it is called 'AMOLF.'

subject—at least in the early years of its existence.[13] Although close relations were maintained with the University of Amsterdam, the laboratory was independent from academia. Most of the research initiated at the lab was relatively short-term, existing for no longer than a few years before being replaced by a new topic.

The second location was the site of a private company, named 'Werkspoor,' which enjoyed a good reputation for construction work; for example, it played a considerable role in the development of the Dutch railway system. Since Werkspoor was located in close proximity to Kistemaker's lab, it was often consulted on engineering matters. When the research on gas centrifuges began, Werkspoor was asked to provide the construction of the centrifuges. The company worked for a relatively low price, in the hope to receive a substantial commission from Kistemaker once gas centrifuges were to be produced on a large scale.[14]

From the beginning precautions were taken to prevent theft and proliferation of knowledge about centrifuges. RCN staff, who comprised approximately half of the staff working on the centrifuge program, were submitted to background checks by both RCN safety officials and the Dutch intelligence service, the 'Binnenlandse Veiligheidsdienst' (BVD).[15] At Kistemaker's lab, a doorman was always present to monitor the movement of staff and visitors. In 1957 the Austrian physicist Gernot Zippe visited the laboratory and shared valuable knowledge on Soviet centrifuge designs with Kistemaker. However, Zippe was not allowed to see the Dutch centrifuges.[16] In 1960, when the FOM laboratory moved to a new location on the outskirts of Amsterdam, a high fence surrounded the new building. Werkspoor was similarly difficult to access for outsiders.

Control of the circulation of technology remained in the hands of the program organizers. RCN, FOM and Kistemaker decided on what could be shared

---

13 Interview by the author with Dolf de Vries, one of the early students of Kistemaker, 21 February 2011; Kistemaker, *De Geschiedenis*.

14 Kistemaker, *De Geschiedenis*, p. 6.

15 I have not been able to trace the exact arrangements on 'screening' for the early centrifuge program. Nonetheless, this point is substantiated by the lists of individuals working at the lab that Kistemaker had to send to RCN's security officer Krips. Furthermore, every RCN worker had to sign a classification agreement before being allowed to work on the centrifuges. See for example: Kistemaker to Krips, 9 July 1956, ET, box 3; "geheimhoudingsovereenkomst" of A. Rijbroek, 19 June 1958, ET, box 3.

16 Gernot Zippe, *Rasende Ofenrohre in stürmischen Zeiten. Ein Erfinderschicksal aus der Geschichte der Uranisotopentrennung im heissen und im kalten Krieg des 20. Jahrhunderts* (Vienna, 2008).

with whom, and what could be published. Whenever suspicions arose about the motivation or integrity of one of the workers, Kistemaker would write to the directors of RCN, or to one of the officials at the Ministry of Economic Affairs, who could intervene if they considered this necessary. One such occasion was Zippe's visit to Kistemaker in 1957. Upon being contacted by Kistemaker, the BVD not only recommended that Kistemaker be careful, but also asked him to gather as much knowledge as possible on the state of affairs in the Soviet Union.[17] For Kistemaker, RCN and Werkspoor this rather unobtrusive procedure was the standard approach. Kistemaker described the safety precautions at the lab actually as "practically non-existent."[18]

An American Request for Secrecy

During the spring of 1960 uranium was separated with gas centrifuges for the first time by Kistemaker and his group. The group working on the project counted approximately 25 individuals. On 11 July 1960 Kistemaker received a letter from the head of the Bureau for Atomic Affairs at the Dutch Ministry of Foreign Affairs, notifying him of a forthcoming visit by a group of American officials who were traveling around Europe. Kistemaker invited them to have lunch at his laboratory on 18 July.[19] The delegation visited Amsterdam for a scientific meeting, and The Hague for a political meeting after having visited Bonn for similar purposes. They had met with officials from the German Ministry of Foreign Affairs and the Ministry of Atomic Affairs, as well as with Wilhelm Groth, one of the leading figures of the German gas centrifuge program. However, no technical details had been discussed.[20]

The delegation consisted of four members. Three worked for the US Atomic Energy Commission (AEC): physicist George A. Kolstad, chief of the physics and mathematics branch of the division of research of the AEC; Algie A. Wells,

17 Willem Reyseger to Jaap Kistemaker, 25 June 1957, ET, box 3.

18 Memorandum written after the visit of the US delegation to Kistemaker's lab, described in the next section, memorandum nr. 1464, from the Adviser for Atomic Affairs (TMA) to his superior, the director-general for European Cooperation (DGES), 21 July 1960, Nationaal Archief, The Hague (hereafter NA), Ministry of Foreign Affairs: Code-Archive 1945–1954, entry number 2.05.117, inventory number 13589. All quotes from Dutch were translated to English by the author.

19 A.N. Baron van Aerssen, head of the Bureau of Atomic Affairs, to Jaap Kistemaker, 11 July 1960, NA, 2.05.117, 13589.

20 Memorandum nr. 1464, TMA to DGES, 21 July 1960, loc. cit.

AEC's director of International Affairs; and Wilbur A. Strauser, deputy director of AEC's division of Classification. The political adviser of the US to the European Atomic Energy Community, Howard Meyers, was the fourth US official on the delegation.[21] The goal of their 'fact-finding' trip, as it was dubbed by Dutch officials,[22] was to consult governments, policy makers and scientists about "the classification of information in the field of ultracentrifuges and the classification policy with regard to its instrumentation."[23] Their primary aim was to come to an agreement about a uniform system for controlling the dissemination of ultracentrifuge knowledge.

The American delegation's petition involved three separate requests. Firstly, they asked that AEC's so-called *Classification guide for the gas centrifuge program* be used as a basis for classification. This two-page document was distributed by the AEC, and designed to provide "guidance" for the categorization of knowledge on gas centrifuges as 'Unclassified,' 'Confidential' or 'Secret.' The guide contained a list of nine different categories of centrifuge properties, most of which were supposed to be classified as secret. Only one point, the "fact of continuing or newly imposed classification of centrifuge work"—meant was whether classified work was done on centrifuges—was labeled as unclassified. The "level of effort on centrifuge program" was to be confidential, and the remainder, mostly technical specifications like the "hydrodynamics of gas flow" and the "information on design," secret.[24] The document itself was classified as confidential.

Secondly, the delegation proposed that the regulations regarding classification be alike across all countries involved. As a first step it was proposed that the Dutch and German government officials intensify their contact, to explore if Germany and the Netherlands would be able to harmonize their classification regime.

Finally, the delegation proposed to classify the entire operation, that is, the act of classifying itself, as secret, despite the fact that this conflicted with

---

21 Report of the meeting directly after the visit of the US delegation between Jaap Kistemaker, W.J. Beekman, J.J. Rijsinge and an official of the Ministry of Foreign Affairs, memorandum nr. 1463, TMA to DGES, 21 July 1960, NA, 2.05.117, 13589.

22 Ibid.

23 Staff member of the Dutch embassy in Washington to the Ministry of Foreign Affairs, 12 July 1960, NA, 2.05.117, 13589.

24 Classification guide for the gas centrifuge program, NA, 'Ministeries voor Algemeene Oorlogvoering van het Koninkrijk (AOK) en van Algemene Zaken (AZ): Kabinet van de Minister-President (KMP), (1924) 1942–1979 (1989),' entry number 2.03.01, inventory number 6701.

the abovementioned sole unclassified aspect of the *Guide*. The US wished to avoid press coverage of the request for classification of the projects. If journalists were to ask questions, the preferred answer would be that this operation merely investigated the possibility of an agreement on peaceful applications of atomic energy; secrecy or gas centrifuges were not to be mentioned. "If the press would find out that it mainly involved ultracentrifuge-experts, this would not be denied, as peaceful (i.e., industrial) goals could be pursued with ultracentrifuges."[25] This remark clearly reflected the main tension accompanying the ultracentrifuge projects: even if this technology was merely applied for peaceful applications, the possibility to enrich weapons-grade uranium with the same technology would always be there.

American Interests: US Control

The origin of the American plea for secrecy can be traced back to the winter of 1959, when Charles L. Marshall, director of the Division of Classification of the AEC, sent a memorandum to Wells, warning against the proliferation of centrifuge knowledge to an "unfriendly nation." Marshall stressed that the technology should not only be classified in the US, but also in Germany and the Netherlands. His advice was to amend existing treaties with West Germany and the Netherlands "to include full cooperation in this field with both nations on a classified basis."[26] This full cooperation would only be possible if the US gained full control over the technology in the Western European countries. Perhaps, he wondered, the US might buy "the fruits of German and Dutch labor?"[27] In the spring of 1960 Wells began working on this scenario. In April he wrote an extensive summarizing report on the topic. The possibility of buying German and Dutch technologies was not pursued, but plans to classify the US program had solidified, as had the aim to try to come to an agreement with the European countries.[28]

25 Memorandum nr. 1463, TMA to DGES, 21 July 1960, loc. cit.

26 Memorandum from Marshall to Wells, 7 December 1959, which was included as "Appendix G" of "Gas Centrifuge Method of Isotope Separation. Report to the General Manager by the Directors of Classification, International Affairs and Research," 9 April 1960, Wilson Centre Digital Archive (hereafter WCDA), accessible at http://digitalarchive.wilsoncenter.org/document/115362 [accessed 28 May 2014].

27 Ibid.

28 "Gas Centrifuge," 9 April 1960, loc. cit.

Concerns about proliferation seemed justified by the latest developments. A relatively recent study by the private Union Carbide Nuclear Company, commissioned by the AEC, had concluded that gas centrifuges—although not yet operational—might provide the easiest and cheapest route towards a uranium bomb, especially for countries without access to large amounts of electricity and material resources.[29] This argument was picked up by the AEC and used by Wells to convince the State Department of its cause: "The capital costs, power requirements, and technical skills necessary to build and operate a production scale plant may shortly be within the capabilities of as many as 20 to 30 foreign countries if development meets expectations and the technology remains unclassified."[30] Classification was necessary to control proliferation.

However, American concerns were not merely, or even primarily, of a military nature. Technological superiority and control also served economic purposes.[31] Since World War II the US had been dominant in the field of uranium enrichment: diffusion plants at Oak Ridge had since been producing large quantities of enriched uranium. Following the introduction of Eisenhower's 'Atoms-for-Peace' policy, this uranium had been sold to other states at a low price. There certainly was a market for enriched uranium, and the US share needed to be secured. Another relatively recent study by General Electric Company had also made clear to the AEC Commission that, within five years, it would be possible to build gas centrifuge plants that would enrich uranium for roughly the same price as that for which the AEC offered uranium to other countries in 1960.[32] Gas centrifuges could be exploited for much profit.

Nevertheless, the AEC realized that the US was not as far ahead in this field as they had thought. Germany and the Netherlands were "vigorously" pursuing their gas centrifuge programs and had, according to Wells, "advanced their technology to the point where it is equal to or better than ours."[33] In particular, the

29 Three possible routes had been compared: gaseous diffusion, gas centrifuges and a plutonium route, via natural uranium reactors. See: S.A. Levin, D.E. Hatch, and E. Von Halle, "Production of Enriched Uranium for Nuclear Weapons by Nations X, Y, and Z by Means of the Gas Centrifuge Process," Operations Analysis Division, Union Carbide Nuclear Company, 26 February 1960, WCDA, accessible at http://digitalarchive.wilsoncenter.org/document/115305 [accessed 28 May 2014].

30 Memorandum for P.J. Farley, 19 February 1960, included as "Appendix D" of "Gas Centrifuge," 9 April 1960, loc. cit.

31 Krige, "US Technological Superiority;" Krige, "Hybrid Knowledge;" Krige, "Maintaining America's Competitive Technological Advantage;" Krige, *American Hegemony*.

32 "Control of and cooperation in gas centrifuge research and development technology," which was "Enclosure to Appendix D" of "Gas Centrifuge," 9 April 1960, loc. cit.

33 Marshall to Wells, 7 December 1959, loc. cit.

German program was considered to be the "most extensive and most complete gas centrifuge program in the world."[34] What should be established, according to the AEC, was bi- or tri-lateral cooperation between the US on the one hand and West Germany and/or the Netherlands on the other, together with classification of the research. In this way, the European technologies would be "controlled" by the AEC, along with being "fully informed of the progress that is being made in this endeavor."[35] Although proliferation was an important argument, securing technological superiority[36]—in order to safeguard the US share of the uranium market, to control European centrifuge technologies, and to anticipate the interest of the US industries—strengthened the need for a request for classification.

Dutch Hopes: Dutch-German Collaboration

The Dutch reaction to the American proposal may be best understood through a consideration of the Dutch hopes and expectations for international cooperation at the time of the request. Of the three other Western nations working on gas centrifuges (West Germany, UK, US), the Germans, and in particular the firm Degussa, seemed to provide the most promising opportunity for cooperation.[37] Degussa was working on the same type of centrifuges as the Dutch, and encountered similar practical problems.[38] Exploratory discussions between the two parties had resulted in far-reaching plans for cooperation.

34 "Gas Centrifuge," 9 April 1960, loc. cit.

35 "General advisory comments and recommendations," included as "Appendix E" to "Gas Centrifuge," 9 April 1960, loc. cit.

36 A term I borrowed from Krige, "US Technological Superiority."

37 Collaboration with the UK was never seriously discussed. The UK had been working on gas centrifuges until 1954, and restarted the work in October 1960 under close collaboration with the US. Before 1960, the Dutch had tried to come to collaboration with the US numerous times, but this was never established, mostly because centrifuge research in the US was classified. R. Scott Kemp, "Gas Centrifuge Theory and Development: A Review of U.S. Programs," *Science & Global Security* 17 (2009), 1–19; R. Scott Kemp, *Nonproliferation Strategy in the Centrifuge Age* (PhD dissertation, Princeton University, 2010), chapter 2. Kistemaker, *De Geschiedenis*, pp. 6–16.

38 As Gernot Zippe had not only shared his knowledge on Soviet centrifuges with the Dutch, but also with Degussa and the US, these two nations were also working on the same type of centrifuges. However, this was probably not known to the Dutch. Gernot Zippe, Jesse W. Beams, A.R. Kuhlthau, *The Development of Short Bowl Ultracentrifuges* (University of Virginia, 1958).

These plans were devised by the director of Degussa, Alfred Boettcher, who personally discussed them with Kistemaker. The latter was very positive about the idea of a collaboration with the Degussa group.

Boettcher's plan for collaboration involved four phases. As neither group knew exactly how far the other group had advanced, cooperation needed to start slowly. The first phase was dedicated to the exchange of patents on gas centrifuges held by the Dutch and Degussa. This would provide both parties with a first impression of the respective projects' status. In the second phase measurement results of the centrifuge performance were to be shared. Thirdly, work programs were to be coordinated. And finally, a real technical cooperation would be formed so that, eventually, all knowledge would be allowed to flow freely between the groups. At that stage the projects would be ready to merge.[39]

While these plans were being formed in the spring of 1960 the question was raised whether this cooperation could be placed within the framework of Euratom. The benefits would include Euratom's financial support for the project as well as the addition of its substantial network for scientific cooperation. Nonetheless, Degussa director Boettcher was strongly opposed to Euratom involvement. His most pressing objection was that all technical knowledge developed within Euratom had to be shared with all Euratom partners, including France and Italy. Boettcher considered this highly problematic, and he also foresaw objections from Euratom's side. Since the French nuclear program was aiming at the development of a nuclear bomb, its use of technical gas centrifuge knowledge might not be confined to peaceful applications; Euratom only supported nuclear cooperation if the goals were peaceful.[40]

Boettcher, who was, according to Kistemaker, "as hard as nails,"[41] added another argument to the opposition against the involvement of Euratom. He was convinced that the development of gas centrifuges could not be successful without American help. In his opinion the plan for cooperation between the three countries would eventually result in contracts between three industrial firms: Degussa, VMF (a conglomerate of Werkspoor and Stork, another Dutch engineering company), and General Electric. In principle, the contracts would 'simply' be bi- or trilateral. The crux of the argument was that, according to Boettcher, the US authorities were not sympathetic towards Euratom, mostly

39 Report of discussions between Kistemaker and Boettcher, FOM-8042, 28 January 1960, NA, 2.03.01, 6701.

40 Report of discussions between Kistemaker, Boettcher and Speicher, 4 January 1960, ET, box 17; Report of discussions, 28 January 1960, loc. cit.

41 Report of discussions, 28 January 1960, loc. cit.

for the reason mentioned above that it required the distribution of knowledge to all countries involved, including France and Italy.[42]

In the spring of 1960 Kistemaker was keen to rush into the first phase of the plan for cooperation between Degussa and the Dutch group. He tried to convince Boettcher that patents ought to be exchanged as soon as possible, to enable each group to familiarize itself with the other group's progress. Boettcher was slow in answering Kistemaker's letters, and seemed reluctant to Kistemaker.[43] No actions were taken before the summer of the same year, and thus not until the American delegation arrived to request that the programs be classified.

Kistemaker's Reaction to the Request

When the delegation visited his laboratory in July 1960, Kistemaker sympathized with the American anxieties about proliferation: he had, in fact, expressed similar concerns to the Ministry of Foreign Affairs just a month previously.[44] According to Kistemaker, the centrifuge route would imply the possibility to construct an atomic "bomb for 30 million" Dutch guilders; this was the estimated cost of the R&D for a production plant.[45] "Any country, no matter how small and underdeveloped, can, by only tweaking the centrifuge a little, make it fit for the production of pure U235, the necessary material for building atomic and H-bombs."[46] Nevertheless, the US request for classification still came as a surprise. When meeting with Kolstad, Kistemaker was asked about the existing Dutch provisions for safety precautions. He answered that "something equivalent to secrecy"[47] was in place. But Kistemaker was bluffing: in the preparations for the meeting he had explicitly requested Dutch officials to present the existing security measures in a light "as positive as possible."[48] In reality, security measures were, as was already mentioned, "practically

42 Ibid.

43 Kistemaker and Veldhuyzen (Werkspoor) to the directors of RCN, 20 April 1960, NA, 2.03.01, 6701.

44 "Instructie ten behoeve van de Nederlandse delegatie naar de multilaterale besprekingen inzake de coördinatie van de classificatie van het Ultracentrifuge-project te Washington," page 1 is missing and the document is not visibly dated, but was probably circulated in February 1962, document number F/2589/62, NA 2.05.117, 13590.

45 Memorandum nr. 1463, TMA to DGES, 21 July 1960, loc. cit.

46 Ibid.

47 Memorandum nr. 1464, TMA to DGES, 21 July 1960, loc. cit.

48 Ibid.

non-existent."[49] A footnote in the minutes reveals that S. Meijer from the Ministry of Foreign Affairs was troubled by, and perhaps a bit ashamed of, this lack of security. He noted that "[s]omething ought to be done about this," and as quickly as possible.[50]

Although Kistemaker understood the American request, he anticipated problems with the implementation of the program's classification, as did RCN and FOM. Kistemaker gathered that neither FOM nor Werkspoor would be keen on classification. Research conducted at FOM was invariably of an academic nature, and performed by individuals who were not familiar with the implementation of secrecy regimes. Besides, Kistemaker's FOM-laboratory also housed projects other than research on gas centrifuges. Radiometric dating, materials research and other, less proliferation sensitive science was done there as well. Classification of the ultracentrifuge program would disconnect it from the rest of the lab, and this was undesirable. When the American classification guide was received at the lab early in August, Kistemaker elaborated on this point. The so-called 'need-to-know' system that the US proposed—which allowed only scientists to access specific types of information, if necessary for their research—would harm the team spirit in his lab to such an extent as to render its influence damaging, according to Kistemaker.[51]

RCN voiced the argument that industry would disapprove of classification of the program. Patents and industrial cooperation had been supposed to form the future basis of the ultracentrifuge project. Philips in Eindhoven (consulted by Kistemaker for practical solutions to engineering problems) and Werkspoor in Amsterdam were already deeply involved in the project, and this involvement was to be expanded in the near future. But how would it be possible to maintain these cooperations if research was classified as secret? According to Werkspoor engineer W. Beets, the government would often pay private firms a special compensation in cases like this, as the technologies could not easily be put on the market.[52]

Another major problem for Kistemaker was that classification of the project would hinder international cooperation significantly. How was it possible to maintain communication with the German groups when nothing could be shared? For Kistemaker, it was highly uncertain if plans for collaboration could

49 Ibid.
50 Ibid.
51 Ibid.
52 Ibid.

be continued within the classified realm. He realized that everything he had tried to build up with Boettcher could be shattered by the American plans.[53]

A few months later, Herman Robert Woltjer, member of the governing council of FOM, entered the discussion and voiced similar concerns about classification, based on the discrepancy between secrecy and the ideal of academic freedom. He expressed concerns about the "screening" of scientists. For him, screening, which would be implemented by the Dutch Secret Service, was out of the question.[54] Apparently, Woltjer was not aware that RCN already had their scientists and engineers screened by the BVD before they started work on gas centrifuges at the FOM laboratory. This rigid stance of the FOM board was not well-received within government circles: these kinds of concerns were irrelevant if "grave dangers for humanity" were involved. Woltjer replied, reluctantly, that he would reconsider the presented option.[55] So, generally, the relevant executive parties (Kistemaker, FOM, RCN, and their industrial partners) knew that proliferation of centrifuge knowledge presented a real threat. However, they had reservations about the American request, based on both practical and ideological arguments connected to the need to share information.

Secrecy for Government Officials

In the months following the visit of the American delegation officials at the Ministries of Foreign Affairs and Economic Affairs started to work on a classification system for the gas centrifuge project. Classifying a certain technology would involve several Ministries. Therefore, a final decision on secrecy had to be taken by the Dutch Cabinet. Officials who worked on this issue and were involved in the subsequent decision making took their task seriously.[56] Most of the coordination was done by officials from the Ministry of Foreign Affairs. Cornelis Fock, Secretary General for the Prime Minister's Office, offered advice on the matter.

Originally, the Ministry of Foreign Affairs interpreted the request mainly from an anti-proliferation perspective.[57] But it was not long before the Dutch started questioning the motives of the US. From whose eyes was the gas

53 Ibid.

54 Memorandum from TMA to DGES, 16 January 1961, NA, 2.05.117, 13590.

55 Ibid.

56 This is witnessed by the high piles of paper still present at the Dutch National Archive on this matter.

57 Memorandum nr. 1463, TMA to DGES, 21 July 1960, loc. cit.

centrifuge to be hidden? It was well known that the Soviet Union had working gas centrifuges in place, so that the American request could not be attributed to concerns grounded in the Cold War conflict. This caused the officials to think that the US wished to hide the technology from other West-European countries, particularly France and Italy.[58]

Another suspicion proposed that the US might be motivated by commercial and economic interests. As ultracentrifuge research was booming in the US, American companies might be the actual parties behind the request. Perhaps the US aimed to delay the German and Dutch programs for as long as possible in order to grant their domestic projects more time for development.[59] Thus, the officials in The Hague were seriously considering the possibility that the US might be using secrecy as a method to secure economic prosperity; as we have seen earlier, this was in fact a correct assessment. It is interesting to note that later in the decade similar concerns were voiced within British government circles on the classification of gas centrifuge knowledge.[60] On both occasions, it was realized that secrecy could be used by the US as a way to secure economic and industrial interests.

The Dutch, however, could not ignore the American request. The proliferation argument would always be valid, whatever other underlying motives the US might have. Nevertheless, officials in The Hague realized that the request could be tailored to fit the specific economic needs of the Dutch project. In addition, hopes for international collaboration were still very much alive and even growing. Some officials believed that the American request for secrecy was only a prelude to a much more comprehensive request for ultracentrifuge cooperation.[61]

Dutch government officials chose to postpone a decision on classification until after Kistemaker had evaluated the *Classification guide for the gas centrifuge program*. His initial reaction, in early September, was that it contained nothing "that would be unrealizable or dishonorable."[62] He did, however, expect "a reduction of possibilities for exchange with other technicians and scientists."[63] As we will see in the following section, Kistemaker tried to make

58 The Dutch Embassy in Bonn to the Ministry of Foreign Affairs, 17 October 1960, NA, 2.05.117, 13589.

59 Ibid.

60 Krige, "Hybrid Knowledge."

61 Memorandum written by Fock, 15 September 1960, NA, 2.05.117, 13589.

62 Kistemaker to Beekman and Rijsinge, written in the beginning of September 1960, NA, 2.05.117, 13589.

63 Ibid.

use of the discussion on classification to argue that the project needed a substantial material impulse.

Kistemaker's Proposal

There were several aspects that Kistemaker wanted to discuss with policy makers in The Hague and at FOM and RCN.[64] These points reflect the impact of secrecy on the involved scientists, and illustrate Kistemaker's concerns about the impending classification. As indicated, he further appears to have hoped for a financial benefit. The fact that he had finally found new, receptive and powerful audiences at the Ministries that operated at a higher level than his usual contacts at FOM and RCN would have helped to tailor his proposal on that point. He tried to render the situation as fruitful as possible. His five most important points will give a comprehensive representation of Kistemaker's argument.

Firstly, classification of the project implied that a new central location would have to be found for centrifuge research as soon as possible. Kistemaker had expressed this wish many times in the preceding years.[65] Since the financing of the project by RCN was never certain for more than a year in advance, Kistemaker hoped that centralization could assure its continuity. A location separate from Werkspoor or Kistemaker's lab would ensure the successful implementation of secrecy; for Kistemaker, this provided a good argument for a new location.

Secondly, Kistemaker was looking for an expansion of the project in several directions. A more solid and expanded financial basis was needed from RCN, as well as closer collaboration, "both organizationally and financially," with industry, in particular with Philips. This, too, was a point that Kistemaker had stressed frequently in previous years. He was hoping to establish funding for the project for more than one year at a time. In addition, he intended to increase the salaries of the scientists and engineers involved in the project. He argued that, if the project were to be classified, the departure of workers should be avoided as much as possible. A stable workforce could, of course, be encouraged by more generous financial remuneration.

A third aspect that Kistemaker wished to change was the level of bureaucracy: RCN demanded extensive and time-consuming quarterly reports.

---

64 Ibid.

65 For example, Kistemaker to Schierbeek, 13 February 1959, ET, box 3; Kistemaker to Boon, 28 May 1959, ET, box 10.

Kistemaker hated this "reportalia."[66] Secrecy would enable to abolish much of it—under classified regulations the technical details for centrifuges would only be accessible to the scientists. Commissioners and board members would have to make do with brief summaries.

A fourth item that troubled Kistemaker was the status of two young scientists working on gas centrifuges for their dissertations: under classification regulations their theses could not be published in the customary channels. Perhaps special arrangements could be made with the universities of Leiden and Amsterdam to withhold the publication of the dissertations when doctorates had been awarded? Or perhaps a kind of "pseudo thesis"[67] could be written in parallel to the actual dissertations, which was then to be made public? The Ministry of Foreign Affairs understood the problem, but nonetheless argued that this was not as pressing a concern as the "responsibility of the Dutch Government regarding this project."[68]

Finally, Kistemaker held that classification would only make sense if the Germans were to follow a similar path. Within a classified realm, contact between the groups could perhaps be maintained, but if different classification rules were applied by Germany, the economic and scientific position of the Dutch program might be jeopardized. Coordination with German policy makers was therefore of the utmost importance, according to Kistemaker. Overall, he recommended that before settling on classification, every bilateral option with West Germany should be investigated.[69]

Press Coverage

On 4 August 1960 an article appeared in *Nucleonics Weekly* with the headline "West Germans agree to put secrecy lid on Gas Centrifuge." The article claimed that "the US has persuaded West Germany to veil in secrecy the gas centrifuge isotope-separation technology under development in that country."[70] Moreover, it pointed to the complicated status (a "government dilemma") of the classification of the gas centrifuge program in West Germany. When classified,

66 Minutes of the Scientific Advisory Commission (WAR) of the ultracentrifuge project, 25 June 1957, ET, box 10.

67 Kistemaker to Beekman and Rijsinge, September 1960, loc. cit.

68 Memorandum from Meijer to Minister President, 6 January 1961, NA, 2.05.117, 13690.

69 Kistemaker to Beekman and Rijsinge, September 1960, loc. cit.

70 "West Germans agree to put secrecy lid on Gas Centrifuge," *Nucleonics Weekly*, 4 August 1960, NA, 2.05.117, 13589.

this could be taken to imply that centrifuges possess a "military importance," which would be a violation of West Germany's post-World War II policy on developing nuclear capability. The country had signed the WEU treaty promising not to produce ABC weapons, but enriching uranium came frightfully close to violating this treaty. Moreover, West Germany was already "psychologically troubled" (as one Dutch government official put it) about the fact that the request for classification might attract international attention to the German program.[71]

The publication in *Nucleonics Weekly* came as a surprise to all. Its assertions were premature. In fact, West Germany had not classified anything yet. The source of the news was a leak from within the US Atomic Energy Commission.[72] The publication led to a stream of coverage on ultracentrifuges in West German media. Degussa released a statement in which it stated firmly that its motives were "peaceful without exception," and that references to cheap atomic weapons, or "A-bombs for the people," were nonsensical.[73]

At the end of September 1960, meetings were organized between the Dutch and West German Ministries of Foreign Affairs. Their aim was to find mutual agreement on the classification of the two gas centrifuge programs. Not only The Hague wanted such an agreement: Bonn, too, took the proliferation argument very seriously and was considering classifying the technology. The recent press coverage accelerated the German decision to classify the gas centrifuge technology as secret; the Germans classified the centrifuge in early October 1960.[74]

Around the same time, the Dutch press began to write about the possible classification of ultracentrifuge research as well. By 13 October practically every Dutch newspaper published articles about the American request to classify ultracentrifuge research.[75] Officials at the Ministry of Foreign Affairs reacted with mild irritation, but no more than that. Within their circles the idea was

71 Memorandum nr. 1463, TMA to DGES, 21 July 1960, loc. cit.

72 The Dutch Embassy in Washington to Ministry of Foreign Affairs, not dated, but probably sent shortly after the Nucleonics publication in the beginning of August, 1960, NA, 2.05.117, 13589.

73 Press release from Degussa, 20 October 1960, ET, box 17.

74 See W.D. Müller, *Geschichte der Kernenergie in der Bundesrepublik Deutschland: Anfänge und Weichenstellungen* (Stuttgart, 1990), p. 504; "Instructie ten behoeve van de Nederlandse delegatie," February 1962, loc. cit.

75 For example: "Duits procédé voor een 'goedkope' kernbom" in *De Tijd/De Maasbode*, 12 October 1960; *Leeuwarder Courant*, 12 October 1960. Both accessible via the Dutch online newspaper depository Delpher [accessed 27 May 2014].

put forward that the article in *Nucleonics Weekly* was published in response to pressure from US industry.[76] Once again, American motives were questioned.

These Dutch publications coincided with a cascade of articles in the Dutch communist newspaper, *De Waarheid*, that kicked off the so-called 'Cellastic affair,' in which Kistemaker was accused of Nazi collaboration during WW II.[77] The fact that he was now actively trying to work together with German scientists, in a field that potentially might lead to the development of an atomic bomb, led communist reporters to accuse him of "German revanchism"—this implied that Kistemaker would be helping old Nazi friends to develop an atomic bomb.[78] The campaign lasted until well into the 1970s. Outside of communist circles, however, these publications were seen as mere expressions of the usual communist slander, in which unconfirmed accusations on both the history and the nature of the centrifuge project were used to vilify Kistemaker.[79] The allegations remained vague, and Kistemaker managed to convince most of the daily newspapers (except *De Waarheid*) to stop writing about ultracentrifuges.[80] In the Netherlands, except for the communist press, there was almost no coverage on the possible classification of the centrifuge project after this point.

By October 1960 every country except for the Netherlands had decided to classify ultracentrifuge research as secret. In the Netherlands, however, the government would not decide on classification until the spring of 1961. In the meantime, the debate had taken yet another unexpected turn.

---

76 Embassy in Bonn to the Ministry of Foreign Affairs, 17 October 1960, NA, 2.05.117, 13589.

77 Samuel Goudsmit, *Alsos : The Failure in German Science* (London, 1947); Rienk Kessenich, *Façades achter een façade. Een bronnenonderzoek naar de Cellastic-affaire* (Master thesis, University of Amsterdam, March 1990); Dirk van Delft, "Preventing Theft: The Kamerlingh Onnes Laboratory in Wartime," in Ad Maas and Hans Hooijmakers (eds.), *Scientific Research in World War II. What Scientists Did in the War* (London, 2009), pp. 62–76.

78 See for example: *De Waarheid*, 23 November 1960, accessible via the Dutch online newspaper depository Delpher [accessed 27 May 2014].

79 See for example: memorandum from TMA to DGES, 28 October 1960, NA, 2.05.117, 13589; summary of *De Waarheid* publications, circulated within government circles and sent to the Minister President, 2 November 1960, NA, 2.03.01, 6701.

80 Kistemaker arranged a meeting with the press. Memorandum from TMA to DGES, 28 October 1960, loc. cit.

Scattered International Dreams

In two meetings held in Bonn and The Hague in October and December 1960, respectively, the Dutch and German delegations discussed the issue of classification, and eventually agreed to use the American *Classification Guide* as a basis for their own guidelines. Nonetheless, both parties expressed their concern about classifying too much technology. Both wanted to keep regulations as narrowly focused on centrifuge technology as possible. The delegations agreed that patents based on research should be filed as secret patents. These would fall under NATO regulations, which would protect the interests of private industry.[81]

But there was an unexpected twist to the discussion: in the Netherlands secret patents could only be filed under the auspices of the Ministry of Defense. This would mean that, despite the peaceful intentions of the program, the centrifuge project would become associated with the military. This would have major implications abroad, as Euratom—still seen as a potential financier—explicitly only supported non-military projects.[82] Furthermore, the Minister of Defense, Simon Hendrik Visser, refused to take responsibility for filing the patents due to budgetary considerations: secret patents normally involved a compensation that was to be paid by the Ministry of Defense to the industries and scientists responsible for the patent. Visser refused to pay for classified centrifuge patents that he had not even requested.[83]

By that time Degussa was considering to cease its gas centrifuge research altogether. The press coverage and the government decision to classify the technology as secret strengthened the idea that Degussa should take a step back.[84] Officials in the Netherlands were not surprised that collaboration with Degussa grew more and more unlikely, until the idea was abandoned altogether in 1961. In the following years, Degussa sold the centrifuge technology to the German government for 5 million German marks.[85] The other German group, under the leadership of Wilhelm Groth, was also gravely troubled by the decision to classify the technology and considered terminating the

81 "Instructie ten behoeve van de Nederlandse delegatie," February 1962, loc. cit.

82 Ibid.

83 Ibid.

84 Memorandum from TMA to DGES, 28 October 1960, loc. cit.; "Oriënterende bespreking in Bonn op 21 en 22 november 1960 door J. Kistemaker," written 28 November 1960, NA, 2.05.117, 13589.

85 Müller, *Geschichte der Kernenergie*, pp. 504–5; Kemp, *Nonproliferation Strategy in the Centrifuge Age*, p. 135.

project, too.[86] Groth changed his mind only after he had been invited to move his research to a new laboratory in Jülich, near Aachen.

Eventually, a few months later (on 10 March 1961), the Dutch government finally decided to classify ultracentrifuge research despite the various drawbacks of this step. It would be classified under military regulations, and henceforth all patents were to be filed as "secret." The centrifuge laboratory was to be transformed into a "secret place." Five patents were directly made secret. In addition, all personnel would be screened and publication of sensitive information henceforth prohibited.[87] A lid was put on the ultracentrifuge.

After the Classification

Once the technology became secret some of the problems that had been foreseen by Kistemaker emerged. The two PhD students' careers were hampered by classification. Joop Los' theoretically oriented thesis on *The separation of heavy isotopes in a centrifugal field* was indeed initially classified as secret.[88] In 1963, Kistemaker wrote a letter to the Ministry of Foreign Affairs, to have it declassified. The Ministry concluded that, as a theoretically oriented thesis, it contained no technical information that was not already in the public domain in 1960. There was no objection against declassification.[89]

Another issue was that the classification regime prevented the contracting out of technical problems to private industry. Before 1960 this had been standard practice. Since private firms were not allowed to learn of some of the characteristics of the apparatus involved, more elements of the centrifuge technology had to be developed at the lab.[90]

Many of Kistemaker's original wishes were granted. A new laboratory was built for the centrifuge project by the Ministry of Economic Affairs and RCN, aided by the fact that FOM requested that the secret project move out of the FOM laboratory by the end of 1961. The financial status of the project improved

86 "Oriënterende bespreking," 28 November 1960, loc. cit.

87 "Instructie ten behoeve van de Nederlandse delegatie," February 1962, loc. cit.

88 Joop Los, *De scheiding van zware isotopen in een centrifugaal veld* (PhD dissertation, Leiden University, 1963).

89 Letter from Jan Hendrik de Boer, chairman of the Scientific Board for Nuclear Energy, circulated within governmental circles, 30 January 1963; Meijer, adviser on atomic affairs for the Ministry of Foreign Affairs, to De Boer, 30 January 1963; Kistemaker to De Boer, 30 January 1963, NA, 2.05.117, 13590.

90 Kistemaker, *De geschiedenis*, pp. 16–8.

considerably. This was partly due to the advice of the so-called 'Tromp Committee,' which advised the Ministry of Economic Affairs on the state of nuclear industrial developments in the Netherlands in 1961. Its assessment of the gas centrifuge project, which also took into account the recent discussion on classification, was strikingly positive.[91] RCN and the ministry decided to expand the project considerably in the first half of the decade, and to provide it secure, long term funding.

However, any hopes for an international agreement dwindled. In early 1961 Boettcher initiated a last discussion, this time to try to form a larger ultracentrifuge "club," together with the UK and the US, within the classified realm. Boettcher was now only willing to collaborate if all four Western countries would work together on the development of the technology, including the US.[92] This demand was new. However, the Dutch Ministry of Foreign Affairs found out that the US had no plans whatsoever for any collaboration following the request for classification. American plans to start a technical cooperation with the Netherlands had been abandoned.[93] In the meantime, the German groups experienced great difficulties because of classification, as described above. These circumstances effectively terminated discussions on international collaboration.

The only remaining Western country working on ultracentrifuges, the UK, had also classified its technology as secret. The UK worked on gas centrifuge research in close collaboration with the US, within the 'special relationship' that existed between the two countries, and with strict classification regimes. Kistemaker had established personal contacts with Hans Kronberger of the Development and Engineering group of the UK Atomic Energy Commission. When asked by Kistemaker to form a "club of the major centrifuge teams," Kronberger replied that under normal circumstances the UK would have welcomed the proposal. "Circumstances, however, are not normal: since military security bans have been imposed on the centrifuge project, international cooperation can only be arranged under treaty provisions. [...] I cannot therefore see how we could collaborate, or exchange information, in the absence of the appropriate treaties."[94] This was secrecy, and thus also American hegemony,

91 "Rapport aangaande het ultra-centrifuge project," written by the Commission for Nuclear Industrial Development for the Ministry of Economic Affairs, July 1961, NA, 2.03.01, 6701.

92 Report of the meeting Kistemaker had with Boettcher, 5 January 1961, NA, 2.05.117, 13590.

93 See for example the letter from Meijer to the Dutch embassy in Washington DC, 12 July 1961, NA, 2.05.117, 13590.

94 Kronberger to Kistemaker, 2 February 1961, NA, 2.05.117, 13590.

in action. Because of the US imposition of arrangements for secrecy on the UK, ties with continental projects could not be established at this point.[95]

The effects of secrecy on Western centrifuge projects implied that the Dutch centrifuge endeavor was forced to go forward alone. Cooperation on the development of an economically feasible uranium enriching ultracentrifuge was not possible during the early 1960s. It would not be until the end of the decade that all parties involved would come to an international agreement on technological cooperation. In 1971 the tripartite Treaty of Almelo was signed by Germany, the UK and the Netherlands, marking the establishment of Urenco, the market leader in uranium enrichment until today. Kistemaker resigned as the director of the program in 1962, at the time when the project moved to the newly built laboratory. He decided that the time was ripe for new leadership, but retained a position as scientific adviser.[96] All technical documents written on gas centrifuge technology after 1 August 1960 are still classified.

Conclusions

A focus on the implementation of secrecy shows clearly that the decision to classify the Dutch ultracentrifuge project as secret was not as straightforward as might be expected. The classification was not simply imposed on the project by the government. On the contrary, both Kistemaker and science organizations FOM and RCN participated in debates preceding the decision. Moreover, most of the recommendations that they voiced on the future of the project were taken over by the Ministry of Economic Affairs.

The case presented here illustrates the relationship between Dutch and American science in the early Cold War. The Dutch simply adapted to the recommendations of the hegemonic US. However, the American motives for urging classification were questioned a number of times: not just the issue of proliferation was at stake, but also the fear that the Americans were protecting their commercial interest in enriched uranium played a major role in Dutch deliberations. One can understand these concerns in particular when considering that the Dutch ultracentrifuge project had been intended to promote the start of a competitive industry. It was unclear to the government how classification would affect that aim, especially as its attainment was considered to require collaboration with Germany.

95 The comparison with Krige, "Hybrid Knowledge," is striking.
96 Kistemaker, *De Geschiedenis*, 1991, pp. 18–9.

The final decision to classify the Dutch project essentially depended on three mutually interdependent developments. Firstly, the fact that the West Germans classified their projects—in part because of the press coverage on the secrecy request—acted as a precedent for the Dutch. Secondly, the subsequent improbability of international collaboration, caused, among other things, by Degussa's decision to stop working on centrifuges, removed at least part of the previous concerns about the difficulties classification would impose on collaboration. Finally, the people that would be effected most by the classification, the involved scientists, offered little resistance to the details of the American proposal, and even recommended that the government officials classify the project. In this way, they seemed prepared to sacrifice their academic freedom for the sake of the common good.

Although scientists could easily be framed as victims of classification (as Kistemaker did in his later writings),[97] this impression cannot be maintained when the subsequent decisions of the Ministry of Economic Affairs are taken into account. Since Kistemaker fully participated in the discussions around the requested classification he could use the situation to his advantage: he could propose certain changes to the project to a powerful and now welcoming forum and thus convince the Ministry of Economic Affairs to invest considerably in the future of the project. Just as the American government used secrecy as a means of implementing 'soft power' in their effort to retain their position in the field of nuclear technologies, Kistemaker used secrecy for his own objectives.

Overall, secrecy proved to have two sides for the Dutch scientists and organizations involved. On the one hand, it produced problems, as it hindered the circulation of knowledge and, in fact, impeded international collaboration. At the same time, it offered leverage to secure a business advantage to the US. On the other hand, a quick-witted scholar like Kistemaker was just as able to enlist secrecy himself as leverage to secure advantages for his project.

---

97 Ibid., p. 16.

CHAPTER 5

Quid Pro Quo: Dutch Defense Research during the Early Cold War

*Jeroen van Dongen and Friso Hoeneveld**

The Netherlands is a small nation that enjoyed a pacifist tradition dating back to the nineteenth century. Soon after World War II, however, it invested substantially in defense-related scientific research.[1] This investment would manifest itself most clearly in the creation of the *Rijksverdedigings-Organisatie* ('National Defense Organization', or 'RVO'). RVO, created on 6 July 1946, brought together existing small-scale laboratories, originally dispersed throughout Holland, with newly created research institutes.

From its inception RVO was a part of the *Nederlandse Organisatie voor Toegepast Natuurkundig Onderzoek* ('the Netherlands Organization for Applied Scientific Research', or 'TNO'), a public and civilian engineering hub, founded in 1932, that would grow into Holland's main institution, for example, for certification and technological consulting. RVO's funding was provided by Holland's Ministry of Defense, in spite of the fact that its institutional parent was TNO. TNO employees, in turn, were not part of the Civil Service, even though most of TNO's operations were funded by the state. Only a limited number of small defense laboratories outside of RVO continued to be subsumed directly under the military; these worked on electronics, aeronautics and ship design. RVO generally chose different subjects and was more geared towards basic science. The placement of RVO within TNO had been a contested outcome of discussions between Holland's military, administrative and scientific elites. Despite

* The work presented here is partly is financed by the Foundation for Fundamental Research on Matter (FOM), which is financially supported by the Netherlands Organization for Scientific Research (NWO).

1 The Dutch budget for defense R&D as percentage of GDP roughly tripled between 1939 and 1948. See the defense budgets for 1939 and 1948: "Rijksbegrooting voor het dienstjaar 1939. 2. VIII. 3. Uitgewerkte en toelichtende staat behoorende bij de raming van uitgaven voor het departement van defensie voor het dienstjaar 1939;" T.H. Rutten, "Memorie van Toelichting, Wijziging van het zesde hoofdstuk der Rijksbegroting voor het dienstjaar 1948, Bijlagen 1136 3 Tweede Kamer," both available at www.statengeneraaldigitaal.nl [accessed 2 April 2014].

 | DOI 10.1163/9789004264229_006

these tensions, most Dutch defense research in the period took place at RVO:[2] it was responsible for roughly 80% of funds spent on defense-related research.[3] These funds grew significantly during the early years of RVO, and beyond: RVO counted 91 staff members in 1947, 236 in 1952, and 559 in 1962.[4]

How can we understand the creation and expansion of RVO? Why would the Netherlands, as a small nation with a traditionally pacifist orientation, choose to invest in military science, especially given the enormous scale at which this was already being developed elsewhere? A perhaps obvious reason why the Netherlands chose to invest in defense-related research is the issue of national security, which was a concern in the period immediately following the German occupation. The important role of science during the recent war gave another incentive to invest in defense research.[5] Yet, these observations do not provide sufficient insight into what the Dutch exactly sought to achieve, or how they wished to reach their goals by creating RVO. Why RVO, with its preference for basic science? How did RVO position itself nationally and internationally? What were its goals? To address these questions we will first provide a short overview of how the expansion of Dutch defense research took shape, with a focus on the creation of RVO and its early years. Central in our subsequent exploration of the abovementioned questions will be the international context in which the Dutch decided to make substantial investments in defense research: scientists, state officials and the military all knew that it was not possible to conduct defense research effectively if its execution was confined to a national scale. We will look at the position the Dutch consciously adopted in light of this observation.

Dutch defense researchers framed their R&D as an integral part of Holland's role in the Western alliance. This resonated particularly well with state officials after the Dutch government had realized in 1951 that direct US economic aid would eventually end, but that American funds could still be attracted via a

2 Gerardus J. Sizoo, "De betekenis van het natuurwetenschappelijk onderzoek ten behoeve van de verdediging in Nederland," *Orgaan van de Vereeniging ter beoefening van de krijgswetenschap* 1 (1959/1960), 29–58; R.H. Kerkhoven, "De betekenis van de Rijksverdedigingsorganisatie TNO," *TNO-nieuws* 27 (1972), 243–6.

3 Budget of Ministry of War, 1957: "Rijksbegroting voor het dienstjaar 1957, Hoofdstuk VIIIA. Oorlog." Available at www.statengeneraaldigitaal.nl [accessed on 1 May 2014].

4 J. Jonker, *Van RVO tot HDO. 40 jaar defensieonderzoek bij TNO* (Delft, 1987), pp. 27, 56.

5 See e.g. Minister of War Alexander H.J.L. Fiévez, in "Installatie van Rijksverdedigingsorganisatie TNO," *TNO-nieuws* 2 (1947), 169–76, on p. 174.

strong domestic military industry.[6] As indicated, state investment in defense R&D was primarily allocated to RVO. RVO took on assignments from the military, but also had its own 'free,' if predominantly defense-oriented, research agenda. RVO was further permitted to farm out tasks to other sectors of TNO. It was also possible for RVO to propose projects to industry. RVO would first conduct the relevant basic research while a contracted company would see through the final product development and, in time, production, for example of some military equipment.

RVO has been the subject of various institutional histories. Its role in chemical weapons research in particular has been studied in great detail.[7] Still, scientists and, in their wake, historians of science have largely and unduly overlooked defense research and the role of RVO in their histories. This applies especially to studies of the socio-political and cultural context of Dutch science during the early Cold War years, and of physics in particular. It is evident, for example, in existing accounts of the life and work of G.J. Sizoo (figure 5.1), who led RVO from 1947 until 1972. Sizoo, a Professor of Physics at the Free University in Amsterdam, was not only RVO's chief executive, but also chairman of its Governing Board.[8] In 1946, prior to his engagement at RVO, Sizoo still expressed an interest in the introduction of an international control of nuclear arms, and he continued to lecture on moral responsibilities of scientists well into the 1950s. While this aspect of Sizoo's work and personality are discussed in the literature, his role in defense research has remained unexplored.[9] A 1994 obituary in the monthly journal of the Dutch Physical Society eschewed all comments on Sizoo's role at RVO and NATO, which easily

---

6 Ine Megens, *American Aid to NATO Allies in the 1950s. The Dutch Case* (Groningen, 1994), pp. 161–95; cf. Till Geiger and Lorenza Sebasta, "National Defense Policies and the Failure of Military Integration in NATO: American Military assistance and Western-European Rearmament 1949–1954," in Francis H. Heller and John R. Gillingham (eds.), *The United States and the Integration of Europe. Legacies of the Postwar Era* (New York, 1996), pp. 253–79.

7 Institutional historiography includes: Kerkhoven, "De betekenis;" Jonker, *Van RVO tot HDO*; Hans Schippers and Harry Lintsen, "TNO en defensie (1946–1990)," in Harry Lintsen (ed.), *Tachtig jaar TNO* (Delft and Eindhoven, 2012), pp. 164–76. On RVO's creation in relation to its chemical work, see Herman Roozenbeek and Jeoffrey van Woensel, *De geest in de fles. De omgang van de Nederlandse defensieorganisatie met chemische strijdmiddelen 1915–1997* (Amsterdam, 2010), pp. 143–73, and Roozenbeek's contribution to the present volume.

8 J. Jonker, "Aan het woord: professor doctor G.J. Sizoo," in Jonker, *Van RVO tot HDO*, 1987, p. 18.

9 See for example: Ab Flipse, *'Hier leert de natuur ons zelf de weg.' Een geschiedenis van de natuurkunde en sterrenkunde aan de VU* (Zoetermeer, 2005), pp. 136–37; Leo Molenaar, *'Wij kunnen het niet langer aan de politici overlaten.' De geschiedenis van het Verbond Wetenschappelijke Onderzoekers 1946–1980* (Delft, 1994), p. 31.

FIGURE 5.1 *G.J. Sizoo, from: C.C. Jonker, "G.J. Sizoo, vijfentwintig jaar Hoogleraar,"* Nederlands Tijdschrift voor Natuurkunde *21 (1955), p. 249.*

constituted the largest part of his professional career;[10] the public picture of a successful physicist was apparently not compatible with his defense research activities. Yet, RVO was clearly shaped by Sizoo to a significant extent, and he was one, perhaps even *the* central figure in Dutch defense research during the early Cold War: apart from his role at RVO, Sizoo was a member of NATO's Science Committee, and he sat on the Board of five research funding bodies in the Netherlands, including a reactor center and the wealthy physics fund 'FOM' (*Stichting Fundamenteel Onderzoek der Materie*, or, 'Foundation for Fundamental Research on Matter'). From 1954 onwards, Sizoo additionally presided over NATO's 'SHAPE' Air Defense Technical Center.[11] SHAPE TC was created in The Hague by an American initiative and fully funded by the US in its initial years. It was designed to conduct research in service of NATO's 'Supreme Headquarters Allied Powers Europe' (or 'SHAPE') in Brussels. A reconsideration of Sizoo and his career, which we will attempt below, will help as we try to understand how defense research in the Netherlands took shape in the early post-war years, and thus address our central concerns.

The Expansion of Post-War Dutch Defense Research

Defense research in the pre-World War II era, and thus in a neutral Holland, operated on a modest scale. The limited number of researchers was essentially divided between just two laboratories, the 'Physical' and the 'Chemical' Laboratories, which were directly under the control of the Ministry of Defense. These laboratories were allocated low budgets, lacked substantial connections to Dutch industry and enjoyed little administrative backing in political and military circles. An international orientation in defense research had not been established at this point; in fact, J.L. van Soest had made only one trip abroad during his tenure as director of the Physical Laboratory before World War II.[12] In 1939, another 'Central' Laboratory was created, but this was, once more, of modest size and had never operated fully when its two researchers, J. van Ormondt and J.H. de Boer, departed for Great Britain upon the German invasion in May 1940.[13]

10 H.J. Boersma, "In memoriam G.J. Sizoo," *Nederlands Tijdschrift voor Natuurkunde* 60 (1994), 44.

11 NATO, Science Committee, Revised List of National Representatives, 19 March 1958, NATO Archives, Brussels, AC/137-N/4.

12 Johannes L. van Soest, *Physisch laboratorium TNO, 1927–1977* (1977), p. 79.

13 See Roozenbeek and van Woensel, *De geest in de fles*.

G.J. Sizoo's early career was perhaps similarly modest in scope, aspiration and execution. Even though he was a full professor, his tenure was at the small, and relatively recent, Free University in Amsterdam; its science department was created in 1930, and Sizoo was installed as its very first Professor of Physics. He was also Holland's first experimental researcher to focus exclusively on nuclear physics, a field that he entered with intent, albeit as a novice, when being appointed professor. It was not until 1935 that Sizoo published his first work on the subject.[14] Soon afterwards, during the years of occupation, physics came to a near-standstill in the Netherlands.

In the case of basic physics, and nuclear physics in particular, the Dutch quickly realized after WWII and the dropping of the atomic bombs in Japan that they needed to substantially enlarge their efforts. At the instigation of physicists like H.A. Kramers and D. Coster, prime minister W. Schermerhorn took the initiative to create the 'Foundation for Fundamental Research on Matter' (or 'FOM'), whose main goal was the creation of a working nuclear reactor. FOM also played the role of coordinating and funding body for all research in physics, and enjoyed a generous budget. Sizoo had been one of FOM's founders, and together with his group at the Free University he benefitted considerably from the FOM budget during the Foundation's earliest years.[15]

Early plans for the reconstitution of, and the substantial investment in, defense research also originated with its practitioners as in the case of basic physics and FOM. Van Ormondt and de Boer drafted several plans for the projected role of defense research in national security. Based on his close observations of the British case, van Ormondt proposed in 1945 to the Minister of War, O.C.A. van Lith de Jeude, that research, including basic science, should play a substantial role in organizing the nation's defenses; he did so even before Holland was liberated.[16] Van Ormondt considered 2% of the ministry's budget

---

14 Abraham Pais, "Kernfysica in Nederland: de beginjaren," *Nederlands Tijdschrift voor Natuurkunde* 57 (1990), 179–85.

15 Friso Hoeneveld and Jeroen van Dongen, "Out of a Clear Blue Sky? FOM, The Bomb and The Boost in Dutch Physics Funding after World War II," *Centaurus* 55 (2013), 264–93.

16 J. van Ormondt, "Organisatie en kosten van research ten bate van de Nederlandsche weermacht," Memorandum 7, 22 February 1945, Nationaal Archief, The Hague (hereafter NA), Ministerie van Defensie te Londen, Ministerie van Oorlog te Londen en afwikkeling daarvan, 1941–1947, entry number 2.13.71, inventory number 3165; ibid., "Memorandum 8", Augustus 1945, Ministry of Defense, Semi-Statische Archiefdiensten, RVO-TNO Archive, Rijswijk (hereafter SSA, RVO), inventory number 891.

an appropriate allocation for the cause.[17] His assessment resonated with the general sentiment of the time, particularly after the occupation had ended and the atomic bombs had been dropped. The first post-war government felt that research in Holland needed to be nurtured and expanded in order to avoid, in the words of Schermerhorn, the demotion of the country into a "Balkan" state.[18] Officials in the Ministry of War had become convinced of the essential role of defense research themselves too, and started to explore its possibilities with TNO as early as on 14 August 1945.[19]

De Boer felt that the civilian TNO organization would be an appropriate choice to locate Dutch defense research. This assessment was primarily based on considerations of cost efficiency: placing defense research with TNO would avoid the accidental duplication of research work, and make special expertise across TNO more readily available to those engaged in defense research. British plans to which de Boer had been privy had shown an inadvertent duplication of research themes and expertise in various departments, an expense he deemed unnecessary and too costly "for the Netherlands," or, rather, for a country of the size of the Netherlands.[20] Within the relatively short period of ca. four months, and immediately following de Boer's report to the minister in February 1946, RVO was created as a subsidiary of TNO.[21]

The positioning of RVO as part of TNO proved to be relevant to its future directions, as it limited the direct influence of the military on RVO's choice of subjects. Indeed, the military criticized an (in its perception) undue emphasis on basic research in RVO's research agenda, and Sizoo was repeatedly forced to defend the RVO organization against military criticism; we will return to this point shortly. When the TNO leadership had appointed Sizoo, one of the qualities they looked for was steadfastness in dealings with the military; significantly, another candidate, the Physical Lab's director van Soest, had been rejected by TNO chairman H.R. Kruyt, as van Soest was perceived as someone

---

17 This amounted to roughly 1 million guilders, which had a purchasing power equivalent to 5 million euro today; Van Ormondt to de Boer, 12 February 1945, NA, 2.13.71, 3165.

18 This was according to Schermerhorn's recollections in the Senate a few years later; see p. 2164 in Proceedings of the Eerste Kamer der Staten Generaal, 14 April 1953, available at www.statengeneraaldigitaal.nl [accessed on 2 April 2014].

19 Minutes of the Meeting of the Executive Board of TNO, 14 August 1945, archive of the Raad van Bestuur TNO, Delft.

20 De Boer to Minister of War Jo Meynen, 28 February 1946, NA, Ministerie van Defensie: Gewoon en Geheim Verbaalarchief en daarbij opgelegde bescheiden, entry number 2.13.151, 6084, sub-inventory 66-S.

21 "Handelingen der Staten Generaal, Rijksbegroting voor het dienstjaar 1948 (Oorlog)" 600 VIII A 8, pp. 37–8, available at www.statengeneraaldigitaal.nl [accessed on 2 April 2014].

who would turn into "a piece of wax" in the hands of senior military.[22] Sizoo had further attractive qualities, including his excellent networks in the academic community (he was chairman of the Dutch Physical Society at the time); his working experience in the laboratory of Philips Electronics, the main industrial lab in Holland; and his amicable relations with government circles (Sizoo had been entrusted with the Dutch Uranium stockpile after the German occupation terminated).[23] The military, however, did not immediately approve of Sizoo's appointment. It was felt that one of its own (colonel S.J. van den Bergh) was equally suited for the job—and that Sizoo would likely present difficulties in the future.[24]

Sizoo's actual influence and areas of control are worth a closer look in this context. In the mid-1950s, RVO consisted of five laboratories. The Physical Laboratory did research on, among other things, digital fire control systems and radar image communication technology. The Chemical Laboratory worked on the detection of, and the protection against, nerve agents, while the Medical-Biological Laboratory mostly concerned itself with detection of, protection against and treatment for damage caused by nuclear and biological weapons. Its emphasis lay on basic research, such as studies into the effects of radioactivity on genes, or the activities of large classes of enzymes. These, together with a Standards Laboratory, were all located near The Hague. RVO also ran an Institute for Sensory Physiology close to Utrecht, near a military air base in Soesterberg. Here, basic and applied research on physical and physiological properties of the human eye were the primary research focus.[25] By Dutch standards these laboratories were well-equipped and successful. They had productive interactions with both industry and academia, and, of course, their military patrons. In 1948 the RVO budget totaled 1.5 million guilders,

22 Kruyt to de Boer, 25 November 1946; de Boer to Kruyt, 2 December 1946, both at Archive J.H. de Boer, Boerhaave Museum, Leiden, 283k.

23 Schermerhorn to Minister of War Jo Meynen, 8 January 1946, NA, Ministeries voor Algemeene Oorlogvoering van het Koninkrijk (AOK) en van Algemene Zaken (AZ): Kabinet van de Minister-President (KMP), entry number 2.03.01, inventory number 6680.

24 Minutes of the Meeting of the Executive Board TNO, 25 February 1947, archive of the Raad van Bestuur TNO, Delft.

25 On the Physical Laboratory, see Van Soest, *Physisch laboratorium TNO*; on the Chemical Laboratory, see Roozenbeek and van Woensel *De geest in de fles*, pp. 405–6, as well as Roozenbeek's contribution to this volume; on the Medical Biological Lab, see A. Querido, "Levensbericht J.A. Cohen," in *Jaarboek der Koninklijke Nederlandse Akademie van Wetenschappen, 1969–1970* (Amsterdam), pp. 261–64; and finally for the Institute for Sensory Physiology, see e.g. M.A. Bouman, "Research Projects at the Institute for Perception RVO-TNO," *Ophthalmologica* 137 (1959), 31.

which was more than double the amount of public funds spent on basic science in Holland (outside of the university system) in the same year.[26] This budget grew even further in the following years. Sizoo and his collaborators thus managed to create a substantial organization in a relatively short period of time: an organization that, overall, appeared to function well. Nevertheless, the arrival of the SHAPE Technical Center on the Physical Laboratory's campus in The Hague may be considered Sizoo's biggest success as RVO chairman. In the following section, we will first discuss how these events developed, with particular attention to the international relations of Dutch defense researchers; and then address the question of why they were perceived to be such a success.

Dutch Defense Research: Motivations and Collaborations

Immediately after World War II the Dutch, particularly former exiles like van Ormondt and de Boer, were intent on retaining contact with British researchers. A delegation led by Kruyt and Sizoo visited Britain in April of 1947, and was well-looked after by Henry Tizard, Chairman of the British Defense Research Policy Committee. Subsequently, contacts with the British intensified, and a sizeable number of RVO researchers traveled to the UK.[27] First attempts to establish contact with the USA were less productive.

After 1949, the willingness of the British to share information with the Dutch was much reduced, most likely because the British did not want to jeopardize their relation with the US by exposing joint work to the Dutch. Tizard voiced his opinion to Sizoo that henceforth, within the Anglo-Dutch collaboration, the Dutch would be best advised to attend primarily to the lesser subjects of food

26 "Handelingen der Staten Generaal, Rijksbegroting voor het dienstjaar 1948 (Oorlog)" 600 VIII A 8, pp. 37–8, available at www.statengeneraaldigitaal.nl. This has the equivalent purchasing power of 6.5 million euro today; for money spent on non-defense related basic research, see Hoeneveld and van Dongen "Out of a clear blue sky?" pp. 264–5.

27 "Rapport over het bezoek van een Nederlandse Militaire Commissie aan Engeland van 15–26 April 1947 in verband met militair-wetenschappelijke samenwerking tussen beide landen," NA, Ministerie van Defensie: Directie Materieel Koninklijke Landmacht (DMKL), voorheen Dienst Kwartiermeester Generaal (DKMG), (1968) 1970–1979 (1981), entry number 2.13.121, inventory number 457; "Verslag over het jaar 1947 van het Physisch Laboratorium van het Ministerie van Oorlog," NA, Ministerie van Marine, sinds 1959 van Defensie (Marine), entry number 2.12.56, inventory number 6555.

and clothing. Furthermore, if RVO staff members wished to visit British facilities in the future, American approval would need to be obtained in advance.[28]

Sizoo decided to travel to the US and meet its senior science officials in person. He managed to arrange a substantial visit for the fall of 1951. During this visit he met, among others, Vannevar Bush, Edward Condon, James Killian, Allan T. Waterman, and general Ward H. Maris, who would later be vital to the implementation of the SHAPE Technical Center in Holland. Sizoo's mission was successful: travel to America by RVO lab directors and their co-workers increased significantly. During these visits some American information was shared with the Dutch, and some American researchers would in turn visit Holland in order to acquaint themselves with Dutch research.

Why would the Americans show an interest in Holland? Sizoo was assured this interest was based on the high level of "fundamental" Dutch defense-related research that had already been attained at RVO.[29] Indeed this impression would be confirmed: within a few years the Americans became so impressed with the quality of RVO's research that they wished to participate in it. Their particular interest lay in RVO's work on fire control and its digital radar-image communication scheme called 'teleplot.' They were further confident of RVO's security regime and liked its embedding through TNO: this made RVO sufficiently independent from any direct state or military meddling. Altogether, these were good reasons for the US to create SHAPE Technical Center on the grounds of RVO's Physical Laboratory. It is interesting to note that RVO also operated SHAPE TC in its early start-up years.[30] It was comparable in size to the Physical Laboratory at the time and dedicated to the development of advance warning air defense systems for NATO.

Contemporary European air defense was complicated: several countries of sizeable population were situated in very close proximity to the USSR. In this close-knit environment, shared air defenses would seem to offer obvious benefits for all relevant parties. However, individual nations, and especially the UK and France, were reluctant to cede command of their air forces to

---

28 Gerardus J. Sizoo, report of meeting with H. Tizard on 13 October 1949, 27 October 1949, NA, Ministerie van Defensie, Generale Staf/Staf van de Bevelhebber der Landstrijdkrachten, entry number 2.13.196, inventory number 3978.

29 Gerardus J. Sizoo to State Secretary of War, 30 December 1954, NA, 2.13.151, 6439; on visits, see e.g. RVO Werkgroep waarneming, report for 1953, archive Instituut voor Zintuigfysiologie TNO, Soesterberg; RVO, "Verslag over het jaar 1952", pp. 3–4, NA, 2.13.196, 3984.

30 F.J. Kranenburg, State Secretary of War, NCRV radio interview with H. Felderhof, 21 January 1955, NA, 2.13.151, 6672; Supreme Headquarters Allied Powers Europe, *The new approach, 1953–1956*, pp. 240–5, NATO Archives, Brussels.

NATO. Additionally, the integration of defense systems and communications produced substantial technical and operational challenges. The US wished to provide a solution to this conundrum by creating a scientific agency—SHAPE TC—that was to provide 'objective' and technical solutions without being perceived as partial to any one nation.[31] SHAPE TC operated with an initial annual budget of about 2.5 million US dollars. Its funding was provided entirely by the US, through its 'Mutual Weapons' development program.[32] The fact that RVO-TNO was not a commercial business was considered an advantage, since it would make the justification of these foreign expenditures to American tax payers easier; the additional benefit of RVO's relative independence from the Dutch government could further help persuade other NATO members of its role with SHAPE TC. Sizoo, who was now greatly appreciated on the other side of the Atlantic, was to serve as SHAPE TC's chairman "at the emphatic request of the Americans," while he continued in the same function with RVO. J. Piket, formerly deputy director of RVO's Physical Laboratory, was appointed as SHAPE TC's first executive director.[33]

Figure 5.2 shows SHAPE TC in 1969; the part of the research campus that has been faded in the image shows RVO's Physical Laboratory. Formally, SHAPE TC and the Physical Laboratory were hardly separate entities in the first few years of the existence of SHAPE TC. RVO was responsible for SHAPE TC's day-to-day running—its salary payments, administrative processes, security, grounds, etcetera—and, at least nominally, also for the execution of its research. A separate foundation had been created as the legal entity that would implement SHAPE's research agenda, but this foundation was run by RVO-TNO, which had been contracted by the US government for this purpose.[34] However, SHAPE TC's staff was recruited internationally, not just in Holland, and its research was more directly aimed at technical consultancy, without the preference for basic science that the RVO labs had.

NATO itself was not initially party to the arrangement between TNO and the US, except that its European Headquarter's research requests were to be

---

31 Raymond H. Dawson and George E. Nicholson, "NATO and the SHAPE Technical Center," *International Organization* 21 (1967), 565–91.

32 See e.g. Gerardus J. Sizoo, "The SHAPE Air Defense Technical Centre," *NATO Letter* 8 (1960), 12–6; Kranenburg, op. cit.; on the SHAPE TC budget, see *Tractatenblad van het Koninkrijk der Nederlanden* nr. 210 (1959), p. 4.

33 Quotation of E.H. van den Beugel, Ministry of Foreign Affairs, to C. Staf, Minister of Defense, 10 December 1954, NA, 2.13.151, 6672; Dawson and Nicholson, "NATO and SHAPE."

34 Cost-reimbursement contract for the establishment and operation of a SHAPE Air Defense Technical Centre, 10 December 1954; "Stichtingsakte SHAPE TC," 15 February 1955, NA, 2.13.151, 6672.

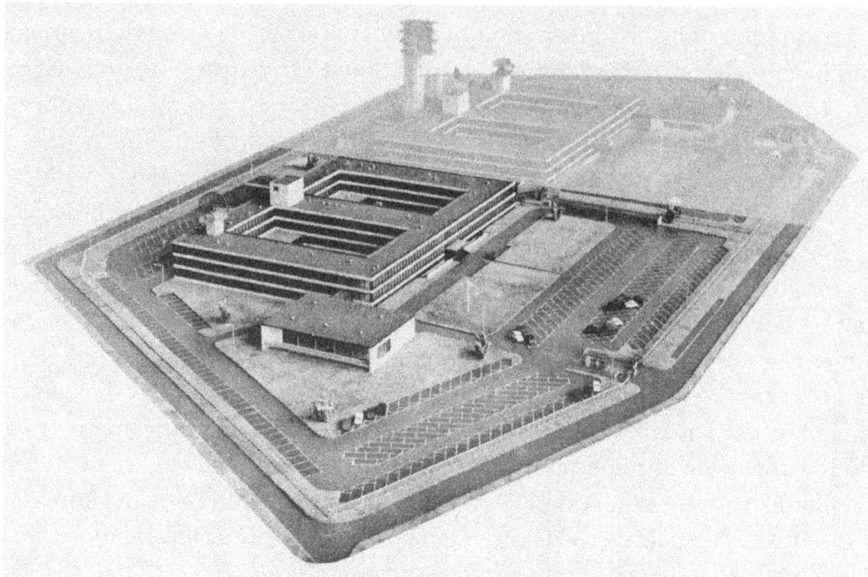

FIGURE 5.2 *SHAPE Technical Center in 1969. In the background (the light part of the photograph) is the Physical Laboratory of RVO.*
SOURCE: *SUPREME ALLIED POWERS EUROPE TECHNICAL CENTER* (THE HAGUE, 1969?).

followed up at the new research center in The Hague. This was greatly appreciated by the NATO organization, and unsurprisingly so. The US clearly expected NATO to take over SHAPE TC as soon as American financial support was going to end, and to put SHAPE TC on its budget. Some British officials initially feared the anticipated costs of this rearrangement, while the French may have entertained the idea to house the Centre itself; but both nations were not asked for input, but simply presented with established facts by the initial American-Dutch agreement.[35] SHAPE TC did transfer to the NATO budget in 1960, while the contractual relationship with RVO, in extension of a prior agreement, was continued until 1963 due to the quality of the work done.[36] Evidently, the addition of a center of this kind was an opportunity too good to disregard for the

35 Van den Beugel, op. cit.; R. Williams to A. MacMahon, 23 December 1954; A. MacMahon to R. Williams, 30 December 1953 (internal correspondence UK Ministry of Defense), The National Archives of the UK, London (hereafter TNA), T225/2478.

36 Dawson and Nicholson, "NATO and the SHAPE Technical Center."

Dutch: its annual turn-over was comparable to the entire budget of the Dutch basic science fund ZWO.[37]

Science policy and related motivations for welcoming the creation of SHAPE TC were very strong indeed, and they explain why its creation may be considered RVO's biggest success. This becomes particularly obvious in light of the Dutch analysis of Holland's international position, and how that perception led the Dutch to invest substantially in defense research throughout the early Cold War years. Dutch researchers realized early, and with some urgency, that they were too minor a player in the international field to maintain independence in defense research. Policy makers, however, were convinced that defense alone, without a research component, was undesirable. Researchers van Ormondt and de Boer stated quite explicitly that international collaboration was a *sine qua non* for any Dutch research effort. Motivated by their experiences in England, they were initially intent on maintaining the fullest form of collaboration with the British, and de Boer informed the Minister of War of this objective on 28 February 1946. The United States were put forward as another possible, and of course desirable, partner for Dutch researchers, while collaboration with continental European countries was not initially considered an attractive option.[38]

Microbiologist J.A. Cohen wrote a report early in 1946 in which he explained his war time activities in England (he had been doing chemical defense research as Dutch reserve officer in Cambridge and Porton Down). It is exemplary for how defense researchers pleaded their case to eagerly listening politicians. Cohen analysed the desired future direction of the Dutch as follows:

> The international role that the Dutch defense organization can play in the future will depend on the contributions that the Netherlands can deliver. In the case of a country like the Netherlands, with small material, yet large scientific resources, it seems evident that these contributions could foremost be in the scientific realm [...]. This requires war research

37 For the ZWO budget see Albert E. Kersten, *Een organisatie van en voor onderzoekers: de Nederlandse Organisatie voor Zuiver-Wetenschappelijk Onderzoek* (Z.W.O.) *1947–1988* (Assen, 1996), p. 120.

38 J. de Boer to Minister of War, 28 February 1946, NA, 2.13.151, 6084, 66-S; J. de Boer to Minister of War, 4 September 1945, NA, Ministerie van Oorlog: Geheim Archief, 1813–1945, entry number 2.13.67, inventory number 327. De Boer held that collaboration with any continental European nation, and in particular Belgium, would lower the Dutch standard. Ironically, van Ormondt would collaborate intensively with Belgian and French researchers on chemical weapons; see the contribution of Roozenbeek in this volume.

> of the highest possible standard [...]. Furthermore, our ability to build our defensive capabilities clearly depends in great measure upon the degree in which we will be introduced to allied technical war secrets. For it to be attractive and rewarding for the allies to come to an exchange with us, it is necessary that war research will again be done at a high level in our own country.[39]

Clearly, Cohen's analysis was not without self-interest, nor indeed without result: Cohen became director of RVO's microbiological lab. But the analysis also resonated with the opinions of other researchers and policy makers,[40] viz. that Holland needed to produce original knowledge if it wanted to be a party to sharing knowledge. Only through access to the knowledge of others would Holland be able to arrive at truly meaningful contributions to the field, which were, in turn, considered essential for securing a prominent enough spot in the Western alliance.

We have already seen that, as high expectations for the partnership with the UK were successively and increasingly frustrated, the USA took on the role of the most desirable partner in this enterprise. John Krige has argued that the US had realized the value of sharing research and its results with partners for the enhancement and securement of American military and technological leadership at an early stage; in some subjects even before 1947. The US considered this practice advantageous even if the boundaries between classified and open knowledge sometimes had to be adjusted for the cause; it would be a vital contribution to the construction of a Western 'arsenal of knowledge' while reinforcing American hegemony on the European continent; at the same time, such interaction would improve American knowledge itself.[41] Nevertheless, not much substantial information was shared before Sizoo's visit in 1951, even though post-WWII American observers were confident about (what they perceived as) a new, serious approach to defense research among the Dutch, who, in their judgment, were forming a "skilled nucleus of

39 J.A. Cohen, "Rapport betreffende researchwerk ten dienste van de chemische oorlogsvoering," after 4 January 1946, SSA, RVO, 887.

40 See for instance J. van Ormondt to Minister of War, 23 October 1945, NA, 2.13.71, 3165; Captain J. Houtsmuller to Colonel S.J. van den Bergh, 13 March 1946, NA, Nederlandse Organisatie voor (Zuiver-)Wetenschappelijk Onderzoek [ZWO], entry number 2.25.36, inventory number 194.

41 John Krige, *American Hegemony and the Postwar Reconstruction of Knowledge of Science in Europe* (Cambridge, MA, 2006); John Krige, "Building the Arsenal of Knowledge," *Centaurus* 52 (2010), 280–96.

scientists and technicians."[42] The reason for the lack of early communication was the relatively little amount of information the Dutch had to offer in the early post-war period—a lack of knowledge that prevented mutual exchange to take place. The Dutch had been warned that this could be the case: Colonel van den Bergh, chief of the military's engineering staff, informed the Minister of War J. Meynen as early as in March of 1946 that the Americans "*appreciate* our cooperation;" yet, they had made very clear that Holland "needs to offer sufficient material in return for what America can tell our country. A one-sided exchange would end very quickly."[43]

After having been forewarned and having arrived at the same analysis themselves, the Dutch had created RVO. This proactive development proved to be beneficial in time: after Sizoo's 1951 visit, American researchers began to share their information on a more substantial scale, and exchange visits were arranged with more frequency and consequence. Sizoo's trip had been intended to establish valuable contacts, but also, and perhaps foremost, to inform the US about RVO and its results, in order to encourage the Americans to share some of their knowledge.[44] Ultimately, investing in RVO enabled the Dutch to engage in *quid pro quo* with the American partner, most prominently through the creation of SHAPE TC in The Hague. As the 1954 annual report of RVO stated, "the nice results concerning radar data transfer obtained by RVO-TNO's Physical Laboratory [i.e., the teleplot technology], along with the willingness of the Dutch authorities to generously share these [...], have contributed in an important way to the decision to establish SHAPE TC in the Netherlands." The Americans benefited from Dutch knowledge as they were co-constructing and securing their hegemonic relation with their novice partner, and in turn, the Dutch gained security, knowledge and substantial financial investment.[45]

---

42 For the positive American assessment of the reinvigorated Dutch approach to defense research, see "Scientific Developments. The Netherlands' postwar CW and BW organization," *Intelligence Review* 33 (1946), 44–49, on p. 49. On the lack of sharing information, see e.g. complaints by Vice Admiraal Termijtelen; minutes of the War Council meeting of 24 May 1946, NA, Ministerraad, entry number 2.02.05.02, 996; on lack of access to information on sonar, see "Verslag over het jaar 1947 van het Physisch laboratorium van het Ministerie van Oorlog," NA, 2.13.121, 506.

43 S.J. van den Bergh to Minister of War, 26 March 1946, NA, 2.13.151, 5899, sub-inventory 06-04/907; emphasis in original.

44 Gerardus J. Sizoo, "Bezoek van een delegatie der Rijksverdedigingsorganisatie T.N.O. aan de Verenigde Staten van Amerika en Canada van 12 September–4 October 1951," NA, 2.13.196, 3979. The official goals for the visit merely listed discussions about the organization of research; the Americans had restricted the subjects to be discussed significantly.

45 See also Kranenburg, op. cit.

Sizoo assessed RVO's work within the context of a potential Dutch contribution to a joint defense strategy within NATO. In his view, excellence of Dutch research was key: "Only someone who has something to offer can receive something, also in exchanges between allies. In the case of science, the quality is of greater importance than the quantity."[46] Exchange was not necessarily meant to imply a literal trade of one idea for another: collaboration was also included. C. Staf, Minister of War and the Navy, further listed the aforementioned possibility to avoid the inadvertent duplication of research as one of the obvious benefits of joint research efforts, as well as the possibility to include more subjects in the alliance. "In the case of a small country such as the Netherlands," Staf explained, "there is a further consideration as well and that is that larger research projects can be embarked upon, if one works in co-operation with other countries, than would be possible if we were to tackle the work alone."[47]

Considerations that lead the Dutch to invest in RVO were not unique to the establishment of that organization: the same argument had played an important role in the decision of the Schermerhorn government to create FOM in 1946. While Dutch physics had been of high quality in the first quarter of the twentieth century, a steady decline set in by the 1930s, which had been accelerated by the German occupation. As indicated earlier, the atomic bombings of Japan made Dutch physicists and their stakeholders in industry and government realize with some urgency that Holland needed to make substantial investments in basic physics, in particular in nuclear science. Here too, a central aim was to create and promote a sufficiently rich and productive scientific culture and infrastructure, so that the Netherlands would be considered a potential partner by other European countries and, of course, the USA.[48] If Holland were not to invest in nuclear science itself, it risked the possibility of being "excluded from this subject in international scientific exchanges,"[49] as Sizoo explicated in 1946 (before he had taken on the leadership of RVO). Policy makers and scientists alike feared to be excluded from the development towards nuclear energy production most. At the same time, it was clear that the Netherlands could not achieve any substantial results without securing

46 Gerardus J. Sizoo, "De betekenis," p. 33. It is not difficult to imagine that RVO's research on fire control would have been of direct relevance for the development of technology to be conducted by SHAPE TC, intended to emit incoming rocket alerts.

47 C. Staf, "The defence departments and the TNO," in *The First Ten Years' Activity of the National Defence Research Council TNO 1947–1957* (The Hague, 1957), p. 9.

48 Hoeneveld and van Dongen, "Out of a Clear Blue Sky?"

49 Gerardus J. Sizoo, "Kernfysica en de Nederlandse wetenschap," *Atoom. Maandblad gewijd aan de atoomenergie en haar gevolgen voor de mens en samenleving* 1 (1946), 14–6, p. 14.

competent collaborators. Dutch physicists and political elites recognized that science was now at the center of international politics, and yet the scale of modern physics prevented the Dutch from an independent, single-nation approach to the field. By investing in its own basic and nuclear physics—J. Kistemaker's isotope separation program[50] and the creation of an accelerator at the new 'Instituut voor Kernfysisch Onderzoek', 'IKO', in Amsterdam, are good examples of this strategic Dutch investment into national physics—the Dutch prepared themselves for admission to the arsenal of knowledge produced and guarded by the US, particularly when the Americans started sharing some nuclear technology after Eisenhower's 1953 'Atoms for Peace' lecture. Earlier, Dutch physicists had already begun working together with Norway on a joint reactor project in the fall of 1951. As in the case of RVO, FOM had been created to engage in the necessary *quid pro quo* of knowledge that was essential to any Dutch role in knowledge production on a large scale in the post-war period. FOM mostly produced open knowledge, while RVO produced 'closed' knowledge, but for both cases analysis and mechanism were the same: the Dutch invested in research to gather sufficient knowledge for mutually constructive collaborations and for *quid pro quo*'s involving sufficiently attractive partners to become possible.

When Sizoo traveled to the opening of the Dutch-Norwegian nuclear reactor at Kjeller, he was alert to the possibility of direct collaborations in defense research with Norway.[51] Sizoo approved of the opportunities that Norway offered, and a little more than two years later, on 16 February 1954, RVO secured an extensive cooperation agreement with its Norwegian counterpart ('Forsvaret forskninginstitutt'), in which "full information [on selected projects]—including 'top secret'—[would] be exchanged."[52] This was a much-appreciated collaboration, for which Sizoo arranged an extension in 1958. Earlier, he had even wished to include it in a tri-partite agreement when the British again wished to collaborate more intensively with RVO. Together with Norway, work on, e.g., guided missiles, radar camouflage and tropospheric scatter communication took place. Notably, the last subject mentioned was of particularly high interest to SHAPE TC.[53]

50 On Kistemaker's research, see Streefland, this volume.

51 Gerardus J. Sizoo, "Verslag van het bezoek van de voorzitter der Rijksverdedigingsorganisatie TNO aan Noorwegen van 28 November-1 December 1951," NA, 2.13.196, 3980.

52 Agreement, RVO-TNO and Forsvaret forskninginstitutt, 16 February 1954, SSA, RVO, 205.

53 Anglo-Dutch Steering Committee for Collaboration on Research and Development, minutes of meeting of 19 June 1958, TNA, DEFE 10/378; Appendix II to 1954 Agreement (see preceding note), 15 September 1958.

The defense research agenda of the Netherlands was influenced significantly by its American orientation. The Dutch reinforced a style of research in which they believed they excelled, and which they believed marked their research as an ideal complement to American projects. Consequently, Dutch defense research maintained a programmatic emphasis on basic science. Indeed, during their 1951 visit, Sizoo and his RVO colleagues found that American research, which was "strongly directed at technology," would benefit from "the more fundamentally oriented 'European mind'."[54] It appears that certain subjects were chosen in the hope that outside interest would be awakened. For example, the Physical Laboratory's program in digital calculation for the benefit of fire control was ambitious from the outset, but would not have been able to succeed without additional, foreign resources; it seems created to attract these. Indeed, the subject did not come to full fruition until the American Mutual Weapons Development program provided additional funding in the 1950s.[55]

In discussions with his military and government patrons, Sizoo stressed the need for RVO to carry out a substantial amount of "free" and "fundamental" research, alongside the work conducted at the direct request of the military.[56] After all, it was this type of research that had attracted, and continued to attract, American attention and contributions in both knowledge and funds, and therefore would benefit the military as well in the long run. The military, however, perceived the situation quite differently. When Major H. Heffener was asked to draft a report on RVO's budget for the Dutch Joint Chiefs of Staff, he was not able to identify a distinction between the activities described by the Medical Biological Laboratory as "fundamental" research and those Heffener himself would consider the personal "hobbies" of its staff.[57] Any existing assessments of, as Sizoo put it, "an international standard" achieved by the laboratory's research and accounts of a great interest from abroad[58] did not overly impress or interest RVO's military patrons, particularly as they felt that

54 Gerardus J. Sizoo, "Bezoek van een delegatie der Rijksverdedigingsorganisatie T.N.O. aan de Verenigde Staten van Amerika en Canada van 12 September–4 October 1951," NA, 2.13.196, 3979, on p. 5.

55 IJ. Boxma, "Digitale rekentechniek voor vuurleidingsproblemen," *Militaire Spectator* 127 (1958), 624–31; van Soest, *Physisch Laboratorium*, pp. 130–7.

56 Rijksverdedigingsorganisatie TNO, "Verslag over het jaar 1952," NA, 2.13.196, 3984; "Verslag van een bespreking inzake het subsidie-beleid RVO-TNO," 1 April 1954, NA, 2.13.151, 6599; Gerardus J. Sizoo, "Taak en werkwijze der Rijksverdedigingsorganisatie TNO," in *15 jaar Rijksverdedigingsorganisatie TNO 1947–1962* (The Hague, 1962), pp. 12–3, 17.

57 H. Heffener, Hoofdkwartier Generale Staf, 29 June 1954, NA, 2.13.196, 3984; see also van Geest, Hoofdkwartier Generale Staf, undated, NA, 2.13.196, 3985.

58 Rijksverdedigingsorganisatie TNO, "Verslag over het jaar 1952," NA, 2.13.196, 3984.

its funding was prioritized over that of more pressing causes. In 1953 frustration reached a point where the Joint Chiefs of Staff wished to create an alternative structure for research and consulting, which was to be under their direct control, and would thus sidestep RVO and TNO. The plan was immediately rejected by Minister Staf, who referred the Joint Chiefs back to Sizoo and his RVO.[59] Evidently, the placement of RVO within TNO aided in securing as much of an independent course as possible, at least when it came to retaining an agenda in basic science and avoiding its direct control through the military.

Conclusion

The scale of Dutch defense research would not obviously define Holland as a 'warfare state'. The UK (its erstwhile desired collaboration partner), or even the similarly modestly sized economy of Sweden would fit this label much more easily.[60] Nevertheless, it has become clear that geopolitical motivations were strongly represented in the Cold War development and growth of the Dutch research infrastructure. While perhaps obvious for the context of defense research, its scale and role in post-war science has been largely overlooked in most existing Dutch historiography. Post-war Dutch science, and physics in particular, has often been presented as internationalist and anti-militarist, in line with physicists' self-representations and the critical attitude that prevailed among them in relation to the military applications of nuclear science.[61] Although FOM was aiming for a peaceful agenda, this cannot be projected to all of post-war Dutch science: the growth and relative size of RVO are necessarily at odds with such an image.

Indeed many Dutch physicists showed more sympathy for non-military research shortly after the war,[62] but this did not obstruct the institution of a visible and accepted role for Sizoo as a prominent physicist both within and outside of the scientific community. Defense research, however, became

59 C. Staf, Minister of War and Marine, to Voorzitter Comité Verenigde Chefs van Staven, 22 June 1953, NA, 2.13.151, 6594.

60 On the UK, see David Edgerton, *Warfare State. Britain 1920–1970* (Cambridge, 2006); on Sweden, see Mikael Nilsson, *Tools of Hegemony. Military Technology and Swedish-American Security Relations, 1945–1962* (Uppsala, 2007).

61 See e.g. Molenaar, *'Wij kunnen het niet langer aan de politici overlaten,'* pp. 71–2, 88–9. On nuclear physics, see Jaap van Splunter, *Kernsplijting en diplomatie. De Nederlandse politiek ten aanzien van de vreedzame toepassing van kernenergie, 1939–1957* (Amsterdam, 1993), p. 109.

62 Molenaar *'Wij kunnen het niet langer aan de politici overlaten.'*

increasingly controversial. In 1965, Sizoo retired as professor[63] in order to dedicate himself more fully to RVO and SHAPE TC. By then activist students perceived RVO's research and its relation with the Americans as highly problematic. In reaction to public criticism, RVO and Sizoo started to justify their role in heated public discussions.[64] While society increasingly problematized defense research, the physics community internalized its own idealist, pacifist self-image. The combination of both factors eventually led to the image reflected in the obituary for Sizoo, in which his role at RVO was entirely omitted.

The creation of RVO and SHAPE TC, as we have emphasized throughout this article, show clearly that Dutch defense research was strengthened after 1945 with a searching eye for attractive collaborative partnerships, particularly with the UK and the US. In 1972, when the military and the RVO leadership reflected on RVO's achievements in its first 25 years, RVO's contributions to original high quality research and the resulting positive effect on the Dutch role in the Western defense alliance were emphasized. Without the expertise of RVO, Holland would not have been able to "keep the flag flying high. In fact, without these contributions, the Netherlands would not even have had access to some essential sources of international information," as a Defense ministry official remarked in 1972.[65] The search for partnerships and collaborations was successful: not only did the desired *quid pro quo* around knowledge actually ensue, the Dutch initial investment also persuaded the Americans to fund an entirely new facility in The Hague, SHAPE TC. Clearly, this greatly facilitated Dutch contributions, through science, to the shared interest of North Atlantic defenses.

Physics fund FOM was similarly created from a desire to recover lost ground in a field of science determined to shape the post-war world. Here, too, the desired partnerships could only be established with immediate and substantial investment at home. In Dutch history of science, much has been written about the expansion of Dutch science during the Schermerhorn administration and why this expansion should be attributed to an ideological belief that a scientifically coordinated society should be created—a society based on principles of rational planning. This perspective proves helpful in under-

---

63 Sizoo did still serve in various administrative functions at the Free University. See H.J. Brinkman, "Toespraak herdenkingsbijeenkomst prof. dr. G.J. Sizoo," 21 January 1994, Archive Sizoo, Free University, Amsterdam, box 5, inventory number 32.

64 See Schippers and Lintsen, "TNO en defensie," p. 175; Gerardus J. Sizoo, "Na vijf en twintig jaar," *TNO-nieuws* 27 (1972), 234–6; Lucas Reijnders, "De militaire geheimen van RVO-TNO," *Vrij Nederland* 29 (7 September 1968), p. 10.

65 Kerkhoven, "De betekenis," p. 245.

standing the foundation of institutions such as the *Centraal Plan Buro* ('CPB', an economics bureau), and the *Mathematisch Centrum* ('MC') in Amsterdam, and even the expansion of the larger TNO organization.[66] Yet the creation of FOM and RVO suggest that the early Cold War context and the desire to find strong Western partnerships in knowledge production were particularly essential to the expansion of Dutch science after 1945, as is further reflected in the disproportionately large budgets of FOM and RVO. The gradual growth of TNO, CPB and MC may be explained as the outcome of a continuous cultural concern that dates back to pre-war years. However, the sudden funding boom that created FOM and RVO are a typical and significant result of their early Cold War context.

66 See e.g. Gerard Alberts, *Jaren van berekening: Toepassingsgerichte initiatieven in de Nederlandse wiskundebeoefening, 1945–1960* (Amsterdam, 1998); Marjan van de Goor, "De wederopbouw van Nederland en de organisatie van wetenschappelijk onderzoek (1945–1947)," *Scientiarum Historia* 26 (2000), 201–16; Gerard Alberts, "Wiskunde en wederopbouw. Deskundigen en hun prometheïsche huiver," *Gewina* 24 (2001), 242–58; Geert Somsen, "Science policy and scientific politics in Britain and the Netherlands: ideas about the planning of science and society in the 1930s and 40s," in K. Bertrams, E. Biémont, G. Vanpaemel and B. van Tiggelen (eds.), *Pour une histoire de la politique scientifique en Europe (XIXe–XXe siècles)*, Mémoires de la Classe des Sciences de l'Académie Royale de Belgique 26 (Brussels, 2007), pp. 77–96.

CHAPTER 6

Chemical Warfare Research in the Netherlands

Herman Roozenbeek

On 23 October 1951, a small group of Dutch military and scientists drove through the Netherlands, Belgium and France to the port of Marseille, where they took a ferry to the North-African port of Algiers. From there, they continued their voyage into the Sahara desert. In charge of the group was J. van Ormondt, the director of the Chemical Laboratory of the defense research organization ('Rijksverdedigings-Organisatie' or 'RVO') of the Netherlands. Their destination was a French testing ground code-named B2-Namous, where the French army conducted secret tests with nerve gases such as tabun and sarin. Apart from the abovementioned two countries, their common neighbor Belgium was also involved. Throughout the 1950s, these countries intimately worked together in the field of nerve gas research, conducting field tests in Algeria, France, Belgium, and the Netherlands. This paper will focus on the Dutch participation in these tests, which remained a closely guarded secret until as recently as five years ago.[1]

Background

Looking back on Dutch military history before World War II, active involvement of the military-scientific community of the Netherlands in advanced, multinational chemical warfare research and field testing was hardly something that was to be expected. Up to 1940 the Netherlands had maintained a strict neutrality and, as a result, had to conduct all its military scientific research in isolation. From 1915 onwards, its armed forces studied the chemical weapons that were first used in World War I, including both research into

1 This paper is based on extensive research in Dutch archives. The author would like to thank the Ministry of Defense of the Netherlands and the Netherlands Organisation for Applied Scientific Research TNO for their kind permission to use documents that were not previously available for historical research. The subject of this paper has been covered at length in: Herman Roozenbeek and Jeoffrey van Woensel, *De geest in de fles. De omgang van de Nederlandse defensieorganisatie met chemische strijdmiddelen 1915–1997* (Amsterdam, 2010). All references to primary sources can be found in this publication.

 | DOI 10.1163/9789004264229_007

protective measures against chemical weapons deployed by an enemy, as well as their active use by Dutch armed forces. The results were, however, quite modest, especially in comparison with those of the great powers that surrounded the Netherlands. Several factors contributed to this outcome, such as Dutch neutrality during the Great War and the resulting lack of experience in combat, the underdeveloped state of the chemical industry in the Netherlands at the time, and the limited budget available to the armed forces in general and to this field of study in particular.

The situation was somewhat different in the Dutch East Indies, especially in the 1930s, when the colonial Royal Netherlands East Indies Army allocated significantly larger resources to counter the chemical warfare threat to the territory from Japan. This lead to the procurement of a small mustard gas installation. Typically, the installation could not be procured from Dutch sources and had to be bought from and built by a German chemical warfare 'entrepreneur' of questionable standards, Anton Cmentek. In the 1920s, for example, Cmentek had operated the installations that provided mustard gas to the Spanish troops fighting the Rif insurgency in Morocco.[2] On the eve of the Japanese attack on the Dutch East Indies, the Royal Netherlands East Indies Army had some stocks of mustard gas available, as well as the necessary equipment for their deployment. However, they were never used, and the supplies remained untouched during the war. The installation was dismantled around 1950, but it was not until 1979 that the stocks were finally destroyed.[3]

A New Orientation

After World War II, the Netherlands abandoned its traditional policy of neutrality. It was one of the founding members of both the Western Union in 1948 and the North Atlantic Treaty Organization (NATO) in 1949. The implications

2 See Rolf-Dieter Müller, "Die deutschen Gaskriegsvorbereitungen 1919–1945. Mit Giftgas zur Weltmacht?" *Militärgeschichtliche Mitteilungen* 27 (1980), 25–54; Rudibert Kunz and Rolf-Dieter Müller, *Giftgas gegen Abd el Krim. Deutschland, Spanien und der Gaskrieg in Spanisch-Marokko 1922–1927* (Einzelschriften zur Militärgeschichte) 34 (Freiburg, 1990); Henning Schweer, *Die Geschichte der Chemischen Fabrik Stoltzenberg bis zum Ende des Zweiten Weltkrieges. Ein Überblick über die Zeit von 1923 bis 1945 unter Einbeziehung des historischen Umfeldes mit einem Ausblick auf die Entwicklung nach 1945* (Diepholz, Stuttgart and Berlin, 2008).

3 M. van Zelm, "Verification of Destruction of CW agents: The Obong operation," in Hans G. Brauch (ed.), *Verification and Arms Control. Implications for European Security. The Results of the Sixth International AFES-PRESS Conference*, 1 (Mosbach, 1990), p. 152.

of this new international orientation extended to the field of military scientific research, and chemical warfare research in particular. The advantages for the Dutch of international cooperation were clear:[4] whereas Dutch military scientific resources always had been too limited to keep up with the current state of military technology in general, the country was capable of concentrating on high-quality fundamental research and sharing the results with military scientists from allied countries. In exchange, Dutch military research would benefit from the research—especially applied research—conducted by international colleagues.[5]

Ideas along these lines took shape in the minds of two Dutch reserve officers, J.H. de Boer and Van Ormondt, both chemical scientists, who happened to work in London during the war. In the 1939 mobilization of the Dutch armed forces, both scientists had been called into active military service as reserve officers to staff a Dutch military chemical laboratory. Their main task was to verify reports about possible German attacks with chemical weapons, but after the outbreak of the war they were sent to England to prevent the Germans from accessing their research papers. In London, they continued their work in support of the allied war effort. There, they learned (nearly) all about the chemical warfare research in both the United Kingdom and the United States. The Dutchmen experienced at first hand what allied military scientific cooperation could accomplish, and learned to appreciate the significance of research coordination. These lessons they would never forget.

After the end of the war, De Boer and Van Ormondt were in a position to apply their wartime experiences to the post-war organization of Dutch military scientific research. They could not have accomplished this without support from leading military circles. Wartime experiences had convinced many in the Dutch military that there was an urgent need "to pay more attention than was done in the past to scientific research," as Van Ormondt observed in a report on the 'Future Development of Military Research in the Netherlands.'[6] This combined effort resulted in the formation of a National Defense Organisation for military scientific research in 1948, under the umbrella of the Netherlands Organisation for Applied Scientific Research TNO, which had been established in 1932.[7] This was a civilian organization which facilitated exchange between

4 See also: Friso Hoeneveld and Jeroen van Dongen, "Out of a Clear Blue Sky? FOM, The Bomb, and The Boost in Dutch Physics Funding after World War II," *Centaurus* 55 (2013), 264–93.

5 See also the contribution by van Dongen and Hoeneveld in this volume.

6 Quoted in Roozenbeek and van Woensel, *De geest in de fles*, p. 157.

7 See also van Dongen and Hoeneveld in this volume.

the military and civilian research communities. Additionally, higher salaries could be offered to scientific personnel here than was possible in the Defense organization.

As a result, and in contrast to most other countries, the Dutch military research organization was not formally part of the Ministry of Defense. Of course, the Defense organization could influence its research agenda and was largely responsible for its funding, but the National Defense Organisation for military scientific research could also spend significant amounts of time and money on any (fundamental) research projects it deemed necessary. Chemical warfare research was concentrated in the Chemical Laboratory in the city of Delft, which was headed by Van Ormondt. A close associate from London, J.A. (Jaap) Cohen—a Jewish scientist from Leiden University who had managed to escape to England during the war to avoid persecution—was appointed director of the associated Medical Biological Laboratory, which conducted pharmacological research on chemical warfare agents. This laboratory also kept the personnel of the Chemical Laboratory under close observation, because working with modern chemical agents carried risks, even with strict safety measures.

The close cooperation with the military helped to provide the research laboratories with scientific personnel. Since 1948 reserve officers with a scientific academic background could be drafted to perform their military service as a research assistant in the military scientific organization. Several of these reserve officers proved to be such a good fit for the research laboratories that they continued to work there after the end of their military service. Among them was A.J.J. (Koos) Ooms, who later became director of the Chemical Laboratory.

Van Ormondt and his associates were convinced that the Netherlands should not only prepare defensive, or passive, measures against possible enemy use of chemical weapons, but also ought to acquire a chemical weapons arsenal of its own. In the Army this position was advocated by a small number of chemical warfare specialists, most notably by lieutenant-colonel F. Sleebos, who as an officer of the Royal Netherlands East Indies Army had played an active part in the extensive chemical warfare preparations in the Dutch East Indies before the war. This was in line with pre-war Dutch doctrine on chemical warfare. In the 1920s and 1930s the Netherlands had not completely ruled out the potential use of chemical weapons by its armed forces in times of war. Although the country had signed and ratified the 1925 Geneva Protocol against chemical and biological warfare, it had done so only with an explicit caveat that permitted the use of these chemical weapons in retaliation, i.e., after an enemy had deployed these weapons first.

After World War II the Dutch government was of a different mind. It did not have any intention to manufacture or deploy chemical weapons under any circumstances. Van Ormondt and Sleebos, who belonged to a small study group on chemical warfare established by the General Staff, repeatedly voiced their own opinions, but without success. By 1948, it was clear to all concerned—although never put down formally until after the end of the Cold War—that the armed forces of the Netherlands would never use chemical weapons themselves. The possible threat of enemy chemical weapons (and by this time, that meant Soviet weapons in particular) had to be met with effective defensive measures only. Incidentally, it should be noted that the Dutch government was convinced—even as late as the 1980s—that circumstances might occur in which the most effective deterrence against any enemy's use of chemical weapons was the certainty that this would be countered by NATO forces using these same type of weapons. The alternative would be to rely on nuclear deterrence alone, and that might lead to an unwelcome escalation of hostilities. In this scenario, however, the Netherlands armed forces would not be required to have access to chemical weapons, because they could rely on the chemical arsenal—the 'chemical umbrella' if you will—of the United States.

Overall, Dutch chemical warfare research focused only on the aspects of prevention, detection, protection, decontamination and treatment. This included basic research on the properties and effects of chemical weapons. From a research point of view, however, there was not much difference between passive-defensive or active-offensive preparations for chemical warfare. Even the development of purely protective measures against chemical agents required detailed knowledge of the properties of all the chemical agents that an enemy could deploy. New toxic chemical substances had to be synthesized and studied, not only in the laboratory, but also in the field. It was, as one Dutch researcher put it, "defensive research in an offensive spirit."[8] Only by approaching the problem from the perspective of an enemy deploying chemical weapons could realistic and effective means of protection be developed.

With respect to chemical warfare research, the post-war years were characterized by a strong sense of urgency. Before and during the war, most experts thought it unlikely that chemical warfare research would come up with powerful new chemical agents. The Dutch military research officers in London, however, did not share this view. De Boer noted that he thought the Germans capable of "astonishing" new developments.[9] He turned out to be right. Only when allied forces entered into Germany, in the spring of 1945, did they become

8 P.A. Jonquière, quoted in Roozenbeek and van Woensel, *De geest in de fles*, p. 239.

9 Quoted in ibid., p. 153.

aware of the existence of German weapons based on newly discovered so-called 'nerve gases,' which were related to common pesticides.[10] Allied intelligence teams discovered secret laboratories, chemical plants and quantities of chemical munitions filled with the first of these nerve gases, tabun. News of other nerve gases quickly followed: sarin and soman. Within days of these discoveries, the Dutch researchers in London were aware of this development and tried to gather as much information as they could. They received valuable information from their allied colleagues, and also received the chance to gather information of their own: Van Ormondt and one of his colleagues were sent to Germany to work as liaison officers to the Chemical Warfare staff of the First Canadian Army.

This phase, however, did not last long. As the end of the war was drawing near, both the British and the Americans were becoming increasingly reluctant to share top secret information about the new weapons with their Dutch ally, mainly because their future military relations to the Netherlands were yet undecided. As long as some Dutch researchers, most notably Cohen, still worked in the United Kingdom—until 1946—they could (without permission, of course) copy classified research papers and other important documents, and hand them over to their colleagues. This well soon dried up. Although Van Ormondt and two of his colleagues were invited to make a tour of the chemical warfare establishments in the United States and Canada in 1947, they had virtually no access to classified research.

This situation was very much regretted by Van Ormondt and his associates. The discovery of the nerve gases had disturbed the balance that had hitherto existed between chemical warfare agents and the defense measures against them. Before new protective gear and procedures could be developed, much research was necessary to determine how nerve gases worked, how they could be deployed in battle, etc. There was, however, another, additional reason for the sense of urgency. The fact that the Germans had been able to hide their successes in the development of nerve gases and the production of nerve gas munitions almost completely from allied intelligence demonstrated to Van Ormondt that, in the future, other totalitarian regimes such as the Soviet Union might also be capable of developing new types of chemical agents and other weaponry in total secrecy.[11] This called for new, more sophisticated methods of scientific intelligence gathering.

---

10 See e.g. Jonathan B. Tucker, *War of Nerves. Chemical Warfare from World War I to Al-Qaeda* (New York, 2006), pp. 70–4 and 84–92.

11 For this failure of allied intelligence, see Reginald V. Jones, *Reflections on Intelligence* (London, 1989), pp. 254–6; Tucker, *War of Nerves*, p. 55.

Since international cooperation in military science was considered vital to the success of Dutch chemical warfare research, Van Ormondt had no choice but to look for other possible partners after the British and Americans had cut off the Dutch. Southern neighbor Belgium was an obvious candidate. In 1946, Van Ormondt visited Brussels, where he was briefed on the research of the Établissement d'Application Chémique, the Belgian counterpart of Van Ormondt's Chemical Laboratory. This lead to a formal agreement to coordinate each country's research and share the results. Much to Van Ormondt's surprise, it soon became evident that the Belgian Établissement had much more experience in the manufacture of nerve gases than his own laboratory. Where and how they had acquired this experience remained unclear.

After the foundation of the Western Union and NATO, the Dutch once more looked to the United Kingdom and the United States for closer cooperation in the field of chemical warfare research, still without much success. Both countries were hesitant to share their expertise with a partner that, at that time, had little to offer in exchange. France, however, was willing to cooperate with the Netherlands and Belgium in a combined research program, which was formalized in 1949. This program included fundamental research into the properties of chemical warfare agents (especially nerve gases), applied research into viable methods of storing and deploying chemical weapons, and the development of means of detection, protection and decontamination. In 1952, a separate bilateral program was agreed upon by France and the Netherlands.

B2-Namous

The 'French connection' soon paid off. In August 1950, at a Western Union chemical warfare meeting, the commander of the French Groupement Arme Chimique invited Van Ormondt to observe the field tests with chemical weapons that the French were conducting in Algeria. The testing ground, the Centre d'Expérimentation Semi-Permanent, was located in a remote part of the Algerian Sahara desert, close to the Wadi (Oued) Namous. It consisted of two parts, a logistics base, B1-Namous, and the testing ground proper, named B2-Namous. Here, the French conducted secret tests with nerve gases. At first they used tabun-filled munitions which they had acquired in occupied Germany at the end of the war. Since 1948, however, French laboratories had been able to manufacture sufficient quantities of nerve gases for testing purposes.[12]

---

12 See: Vincent Jauvert, "Quand la France testait des armes chimiques en Algérie," *Le Nouvel Observateur*, 23–29 October 1997; Claude Meyer, *L'arme chimique* (Paris, 2001), pp. 154–61.

Two Dutch researchers, Van Ormondt and Sleebos, visited B2-Namous in November 1950. Because of the weather conditions this was the preferred time of year. In four days, they witnessed four experiments, each involving 240 nerve gas grenades. The Dutch were vastly impressed, but also highly critical of the scientific quality and value of the tests. The French research methods were, in their opinion, crude at best. Van Ormondt shared his opinion with his French colleague, suggesting that better measurement equipment would yield more reliable results. He also made clear that he was anxious to actively participate in next year's tests. Large-scale tests such as these could not be carried out in the Netherlands. The French reacted favorably to this, and formally invited the Dutch researchers for the 1951 field tests. The Belgian colleagues also received an invitation. Together, they were able to improve the scientific value of the tests by using more and better measurement and analysis equipment. In this way, it was possible to get a more accurate and complete picture of the contamination effects.

In order to test the new equipment and working procedures, two small field tests were conducted in the Netherlands. The French provided a few kilograms of tabun for these purposes. In the Sahara, the larger-scale tests also included sarin, provided by the Belgians, in order to compare the properties and effects of both nerve gases. Despite all preparations, the field tests of 1951, which included aerial bombardments with tabun-filled munitions, were not entirely successful. Due to adverse weather conditions and technical difficulties, many experiments had to be cancelled. A prototype of a detection device for nerve gases, developed by Sleebos, for example, had been damaged during the long trip to the Sahara and failed to work. Nevertheless, the tests suggested that sarin was even more effective than tabun.

The researchers were confident that future field tests would provide more and better results. To that end, the three countries put more effort into the planning of the experiments and to standardization of equipment. The Dutch researchers focused mainly on developing more accurate methods of detecting and measuring nerve gas diffusion, especially at larger distances from the detonation point. Although the General Staff of the Royal Netherlands Army questioned the relevance of these large-scale field tests for the comparatively limited Dutch purposes, Van Ormondt could continue on his own without much interference for the time being. In his report, Sleebos duly pointed out that participation in these tests was vital to the development of effective protective measures, and that it allowed the scientists to validate laboratory observations under realistic conditions. He also noted the relevance of large-scale field tests for the development of chemical weapons, even though that argument did not apply to the Dutch situation.

Preceded by another series of small-scale tests in the Netherlands, the 1952 field tests in Algeria were much more successful than those of the previous year. Next to the French, Belgians and Dutch, smaller delegations from the United States and the United Kingdom also participated. The experiments now showed convincingly that sarin was more effective than tabun. Other nerve gases were also evaluated. This included a new variant of soman which had been discovered by Dutch researchers of the Chemical Laboratory only recently. This substance, code-named 'Agent X,' proved to be very toxic, but highly unstable, and therefore not suited for military use.

Since the next series of field tests would require much larger amounts of sarin—as much as 2,000 kilograms—, plans for 1953 were postponed. This delay also provided a useful interlude in which to improve the quality of the techniques employed to determine nerve gas contamination further. In the meantime, it became increasingly difficult for Van Ormondt to receive approval for his plans. Captain G.M.M. Houben, who acted as representative of the Dutch General Staff at the 1952 field tests, had written an unfavorable report: he was very critical of the French and Belgian research methods and concluded that the Dutch team's assistance had been indispensable for many of the experiments to yield meaningful results at all. According to Houben, the American delegation had made the same observations. More pertinently, he criticized that not enough attention was given to the priorities of the Dutch military, which lay in developing protective equipment. Van Ormondt, he believed, was more interested in fundamental scientific research than in practical questions. Incidentally, Sleebos agreed with this view.

Houben also raised more fundamental objections to Dutch participation. Discussions with French colleagues in Algeria had led him to believe that their main interest was the development of an arsenal of French chemical weapons.[13] "It strikes me as odd," he remarked, "that Dutch help is used to ascertain which nerve gas is best suited for this purpose. And even more when it comes to determining what type of chemical grenade is to be used."[14] In short, the French were profiting more from Dutch cooperation than vice versa. However, Van Ormondt and his associates did not seem to be concerned about this. They focused on their own research and believed the field tests in Algeria to be of vital importance to Dutch purposes. Only by cooperating closely with the French could the Dutch gain access to the Algerian testing ground. The Dutch also had little experience in manufacturing large quantities of nerve gas, expertise which the French and the Belgians could provide. Gradually, however, the Dutch General Staff was able to claim more influence

---

13 Cf. Meyer, *L'arme chimique*, p. 159; Olivier Lepick, *Les armes chimiques* (Paris, 1999), pp. 97–8.

14 Quoted in Roozenbeek and van Woensel, *De geest in de fles*, p. 198.

over the research agenda, which meant that more attention was to be given to the development of effective procedures for the decontamination of military equipment.

The field tests in Algeria scheduled for 1954 were also postponed several years in a row. Apart from the aforementioned reasons two external developments contributed to this. First, the French nuclear weapons program was given priority, which limited resources for the chemical warfare program. Second, the Algerian War of Independence starting in 1954 complicated the organization of the field tests. Nevertheless, the tests continued intermittently throughout the war. In 1956 the French and Belgians conducted experiments on a smaller scale together. The Dutch preferred not to participate, as they did not expect any groundbreaking results from the French-Belgian test series. This proved to be something akin to a self-fulfilling prophecy: the absence of the Dutch researchers and their equipment was one of the reasons for the experiments' lack of success.

On a smaller scale, the three countries continued to work closely together during field tests with chemical warfare agents in continental Europe. Testing grounds were set up at Elsenborn in Belgium and at Mourmelon in France. The three countries also sought to improve contacts with much more experienced colleagues from the United States, in the expectation that this would lead to a mutually beneficial relationship—an erroneous expectation, as would soon transpire. The Americans kept a close eye on the chemical warfare exploits of their continental European allies, but were not generally prepared to grant them insight into their own research. This frustrated the French in particular.

After years of preparation, the large-scale field tests in the Sahara were resumed in 1958. A few months previously, all procedures had been drilled to perfection at Mourmelon. Tests were conducted with grenades and rockets, filled with sarin. To deploy these munitions, twelve guns and two multiple rocket launchers were available to fire simultaneously. Although the weather conditions caused, again, the cancellation of several experiments, the 1958 field tests may definitely be considered the high point of the French-Belgian-Dutch chemical warfare cooperation. This was also the last time that the Netherlands actively participated in the series. The Dutch declined an invitation for the field tests of the following year because of the war in Algeria, and for unknown reasons never received another invitation. Nevertheless, they still worked together with French and Belgian colleagues during the smaller-scale field tests at Mourmelon and elsewhere, at least until 1964. In this way, they still contributed to the success of the Sahara field tests, albeit indirectly.

Although Van Ormondt later declared that the Netherlands had not made any contribution to the chemical weapons programs of their major allies, this was far from the truth. Dutch participation in the field tests with nerve gases certainly helped the French to obtain valuable scientific information about the effectiveness of the chemical-weapons prototypes tested by them. The border between active and passive aspects of chemical warfare research had proved to be extremely thin.

Military Scientific Intelligence

Since 1952, participation in the Sahara field tests had raised the international reputation of Dutch chemical warfare researchers to such a degree that they received invitations from both the Americans and the British to visit their own testing grounds for chemical warfare; these invitations were for visits to the US Army Chemical Center in Edgewood and the Chemical Defense Experimental Establishment in Porton Down respectively. Van Ormondt apparently considered this to be one of the most important results of the French-Belgian-Dutch collaboration effort. Since his close cooperation with the British and Americans during the war, he had always held them in high regard. Cohen, the director of the Medical Biological Laboratory, concluded after the visit to the United States that the choice to focus on high-quality fundamental research had finally paid off. The Dutch scientists now had something to offer their American colleagues, and in return they received valuable information which they would never have been able to gather on their own. Starting in the 1960s, the Americans also contracted chemical warfare research projects in the Netherlands, which, in time, made the Dutch military scientific research organization less dependent on funds from Holland's Department of Defense. Despite their admiration for American achievements, Van Ormondt and Cohen were critical of the Americans' exclusive focus on active-offensive research. By neglecting research into substances that were not relevant to their own weapons program, the Americans ran the risk of being unprepared for the possible use of these chemical agents by the Soviet forces, the Dutch argued.

The closer relations with British chemical warfare research were also beneficial. In 1957, the British informed their Dutch colleagues—before they informed several other NATO-partners, and under the promise of complete secrecy—of the discovery of so-called 'V-agents.' This was a new, even more lethal family of nerve gases discovered almost by accident in the United Kingdom a few years earlier.[15] This information enabled Dutch researchers to synthesize such substances themselves and, within a year, manufacture sufficient quantities for

15 Tucker, *War of Nerves*, pp. 146–7, 154 and 158–60.

laboratory research. The study of V-agents—especially the VX-variant—was given priority over other projects, not least because a NATO intelligence report known to Van Ormondt had warned that the Soviet Union had also learned of the existence of V-agents. The report estimated that the Soviet armed forces would be able to deploy chemical munitions based on V-agents by 1960. (We now know, however, that the Soviets were not able to field V-agent munitions until several years later.)[16]

In general, Van Ormondt valued reliable military scientific intelligence very highly. His wartime experiences in London had certainly contributed to this. He and his colleagues developed close ties to the Dutch military intelligence community. The connection worked two ways: the scientists received reports on chemical warfare developments in the Soviet Union and in other countries, and in turn they commented upon these reports, evaluated their reliability and advised the military on their implications. Not all of these reports were insightful, but at times they included invaluable bits of information. These exchanges were not limited to the Netherlands: Van Ormondt also cultivated and expanded his relations with British and American military scientific intelligence. On several occasions, Dutch scientists were briefed by American intelligence officers on the chemical warfare capabilities of the Soviet Union and its Warsaw Pact satellites. Although this may imply that the Americans valued the feedback from their Dutch colleagues, it also seems likely that they were feeding their allies with reports on Soviet capabilities in order to convince them that, even in the nuclear age, chemical warfare was still a real possibility, and that their armed forces should be prepared for it. Van Ormondt, his colleagues in research and the Dutch military did not need much convincing.

Farewell to B2-Namous

The French tests in the Sahara desert continued even after the independence of Algeria in 1962. A secret clause to the treaty of Évian granted the French access to the testing ground for another five years. In 1965, a French invitation brought the Dutch back to the testing ground at B2-Namous, but only for observation, not participation. Van Ormondt, in his last year as director of the Chemical Laboratory, did not hesitate to accept. For the first time, the French would conduct tests with V-agents. After years of research and experimentation, the French were now capable of producing VX in sufficient quantities for large-scale field tests. Even in a role of mere observers, the Dutch hoped to obtain

16 Ibid., pp. 180–2.

useful information about the properties of VX. One Dutch scientist could note in his report that VX clearly was "a powerful anti-personnel weapon."[17]

In 1966 the Dutch returned once more to the Algerian testing ground. This was mostly a farewell visit, since the French rights to use the establishment were about to expire. A Dutch offer to participate actively again had been rejected by the French. For the occasion, the delegation from the Netherlands consisted of Van Ormondt, his successor A.A.J. Ooms and an officer representing the Army. The three noted in their report that the French tests had been successful to such a degree that they were ready to produce VX-munitions if they chose to do so. The Dutch delegation also managed to gather much-needed information on the properties of VX-contamination under realistic circumstances, which was of vital use to their research on protective measures. The scheduled termination of the field tests in Algeria was much regretted in the Netherlands, since the research on V-agents was considered to be still in its infancy, and Dutch scientists were hoping to get more out of their cooperation with the French. In the end, the secret arrangement between France and Algeria concerning B2-Namous was extended by another five years, but the Dutch were not welcome any more, not even as observers. The new French attitude towards NATO may have been responsible for this.

By the time of the farewell visit to Algeria, Dutch chemical warfare research was ready for a reorientation. For the Netherlands, which focused on so-called 'passive' aspects of chemical warfare such as detection, protection and decontamination, large-scale tests such as those in the Algerian desert were not always necessary. In the 1960s, partly due to the greater influence of the General Staff on chemical warfare research, decontamination of military vehicles and equipment became the focus of Dutch research. A large group of NATO-countries worked together on this subject. Research data from the Algerian field tests enabled Dutch researchers to construct and validate a mathematical model of the dispersion of nerve gases, making future field tests, to some extent, redundant. For the Netherlands, this venture into a working mathematical model proved to be the most important legacy of the Sahara field tests: because of this research close ties developed, which exist up to this day, between Dutch and American chemical warfare research.

Under its new director Ooms the Chemical Laboratory became increasingly involved in initiatives to develop a new, better chemical weapons treaty to replace the ineffective 1925 Geneva Protocol. It was only after the end of the Cold War, in 1993, that the Chemical Weapons Convention (CWC) could finally be concluded; it entered into force in 1997. The Dutch researchers'

17 Quoted in Roozenbeek and van Woensel, *De geest in de fles*, p. 257.

development of effective verification methods for monitoring compliance with this convention benefitted in no small degree from the data collected at the field tests in Algeria. The Dutch involvement contributed to the selection, in 1992, of The Hague as the seat of the Organisation for the Prohibition of Chemical Weapons (OPCW), which was founded to administer the CWC.

Conclusion

Over the years, Dutch chemical warfare research profited much from the close collaboration with French and Belgian colleagues. Dutch researchers participated in three series of field tests in Algeria in the 1950s and were present as observers a few times afterwards. Participation in smaller-scale field tests at Mourmelon and elsewhere was almost as important. This type of close cooperation proved to be more rewarding than the more casual relations within larger organizations such as NATO. The international orientation advocated by Van Ormondt, and facilitated by Cold War conditions, had changed the Netherlands from a minor player in chemical warfare pre-World War II, into a sought-after partner for international military scientific cooperation up to this day.

CHAPTER 7

The Fulbright Program in the Netherlands: An Example of Science Diplomacy

Giles Scott-Smith

Educational exchange continues to be a matter of importance for the United States. At his first speaking engagement as US Secretary of State in February 2013, John Kerry quoted William J. Fulbright, initiator of the famous international educational exchange program: "Senator Fulbright [...] knew that the value of sharing our proudest values bore fruit in the long run [...]. 'Having people who understand your thought [...] is much greater security than another submarine'."[1] The Program was initiated by Senator William J. Fulbright for several reasons: to reduce international tensions and promote peace through international understanding; to extend US influence abroad, and enable Americans to better understand other nations (and their own place in the world); to inform future leaders about and orientate them on the United States; and finally to generate public-private cooperation in furthering these goals. This chapter investigates the role and influence of the Fulbright Program, the principal mechanism for promoting educational interchange into and out of the United States, with a focus on the Netherlands in the period between 1949 and 1980. It concentrates on the impact of the Program across all academic disciplines, but with special attention for the hard sciences.

The Program was a flexible tool that could be adjusted to meet the needs of both the American and Dutch governments. Particular institutions—most notably the Dutch Foundation for Fundamental Research of Matter (FOM)—were able to make use of the Fulbright Program's benefits to further their specific agendas in promoting scientific research in the Netherlands. Fulbright became a prime mechanism for making ideas travel, and this required individuals to carry ideas: as Robert Oppenheimer commented, "perhaps the best way to send knowledge is to wrap it up in a person."[2]

1 John Kerry, "Address at the University of Virginia," Charlottesville, Virginia, 20 February 2013, available at http://www.state.gov/secretary/remarks/2013/02/205021.htm [accessed 18 March 2014].

2 Quoted in: Institute of International Education, US Department of State, Office of Educational Exchange, *Building Roads to Peace: Exchange of People between the United States and Other Countries* (New York, 1948).

 | DOI 10.1163/9789004264229_008

The Fulbright Program has long been presented in the guise of US 'benevolent globalism.' The sale of surplus war materials and the channeling of their proceeds into funding international educational exchange has been put forward as the epitome of enlightened superpower leadership since its inception. Randall Woods described this approach as "an integral part of the internationalist movement that swept America during and after World War II," and according to him, the Program expressed a "mission [...] to make the world safe for diversity."[3] Yet, these insights have not been applied to in-depth research on the Program's impact. A recent study on the origins of US public diplomacy during and after WWII refers only in passing to the merit of the Fulbright Program for a "mutual exchange of ideas."[4] The only book-length study of the Program dates from 1965, followed by several collections of alumni reminiscences and historical essays.[5] Finally, scholars examining the expansion of the US culture and information campaigns during the Cold War have tended to subsume Fulbright under this broader ideological offensive too easily, losing sight of its unique aspects and operational mentality.[6]

Historians are now re-examining the Program's origins, interpreting it in the frame of 'nationalist globalism' as opposed to the dominant discourse of either benevolent hegemony or its critique, cultural imperialism.[7] Studies on individual countries, which follow the practice of the Program closely in particular cultural and political settings, are now breaking new ground to question its 'apolitical' status and highlight the long-term impact of the relationships that have been built up over time. So far Australia, Austria, Benin, China, and Spain have been covered, either entirely or in part.[8] In the late 1990s, the late

3 Randall Bennett Woods, "Fulbright Internationalism," *Annals of the American Academy of Political and Social Science* 491 (1987), 22–35, on p. 23.

4 Justin Hart, *Empire of Ideas: The Origins of Public Diplomacy and the Transformation of US Foreign Policy* (Oxford, 2013), p. 6.

5 Walter Johnson and Francis Colligan, *The Fulbright Program: A History* (Chicago, 1965); Arthur Power Dudden and Russell Dynes (eds.), *The Fulbright Experience 1946–1986: Encounters and Transformations* (New Brunswick, NJ, 1986); Leonard Sussman, *The Culture of Freedom: The Small World of Fulbright Scholars* (Lanham, MD, 1992); Richard Arndt and David Lee Rubin (eds.), *The Fulbright Difference* (New Brunswick, NJ, 1996).

6 See Walter Hixson, *Parting the Curtain: Propaganda, Culture and the Cold War 1945–1961* (New York, 1997); Liping Bu, *Making the World Like Us: Education, Cultural Expansion and the American Century* (Westport, CT, 2002).

7 Sam Lebovic, "From War Junk to Educational Exchange: The World War II Origins of the Fulbright Program and the Foundations of American Cultural Globalism, 1945–1950," *Diplomatic History* 37 (2013), 280–312.

8 Frank Salamone, *The Fulbright Experience in Benin* (Williamsburg, VA, 1994); Guangqiu Xu, "The Ideological and Political Impact of US Fulbrighters on Chinese Students: 1979–1989,"

Jan Rupp of the University of Amsterdam carried out research on the Fulbright Program in the Netherlands, but was unable to complete his study.[9] The contribution of Fulbright to the development of specific academic fields such as American Studies has been studied in some depth.[10] Social scientific methods have been introduced to test, both qualitatively and quantitatively, the impact of exchanges on specific groups of participants.[11] Broader frameworks for interpreting and analyzing educational exchange within the context of US global power, specifically in the fields of knowledge production and international education, have also been put forward, by Paul Kramer and David Engerman.[12] This chapter contributes to this school of studies by examining in detail the Program's role in bilateral knowledge transfer between the Netherlands and the United States, in the first decade of its existence. It thereby explores the ways in which the program was a process of negotiation (rather than a simple one-way transfer of US expertise to a smaller allied nation), and shows how the Netherlands made strategic use of the Program's opportunities to raise its

---

Asian Affairs 26 (1999), 139–57; Thomas König, *Das Fulbright in Wien: Wissenschaftspolitik und Sozialwissenschaften am 'versunkenen Kontinent'* (PhD dissertation, University of Vienna, 2008); Lorenzo Delgado Gomez-Escalonilla, *Westerly Wind: The Fulbright Program in Spain* (Madrid, 2009); Matt Loayza, "A Curative and Creative Force: The Exchange of Persons Program and Eisenhower's Inter-American Policies 1953–1961," *Diplomatic History* 37 (2013), 946–70; Alice Garner and Diane Kirby, "'Never a Machine for Propaganda?' The Australian-American Fulbright Program and Australia's Cold War," *Australian Historical Studies* 44 (2013), 117–33.

9 See Jan C.C. Rupp, "The Fulbright Program, or the Surplus Value of Officially Organized Academic Exchange," *Journal of Studies in International Education* 3 (1999), 59–82; Jan C.C. Rupp, *Van oude en nieuwe universiteiten: de verdringing van Duitse door Amerikaanse invloeden op de wetenschapsbeoefening en het hoger onderwijs in Nederland, 1945–1995* (The Hague, 1997).

10 Giles Scott-Smith, "The Ties That Bind: Dutch-American Relations, US Public Diplomacy and the Promotion of American Studies in the Netherlands since the Second World War," *The Hague Journal of Diplomacy* 2 (2007), 283–305; K.H. Füssl, "Between Elitism and Educational Reform: German-American Exchange Programs," in D. Junker (ed.), *The United States and Germany in the Era of the Cold War, 1945–1990*, Vol. 1 (Cambridge, 2004).

11 See Carol Atkinson, "Does Soft Power Matter? A Comparative Analysis of Student Exchange Programs 1980–2006," *Foreign Policy Analysis* 6 (2010), 1–22; Ian Wilson, *International Education Programs and Political Influence: Manufacturing Sympathy?* (Basingstoke, 2014).

12 David Engerman, "American Knowledge and Global Power," *Diplomatic History* 31 (2007), 599–622; Paul Kramer, "Is the World Our Campus? International Students and US Global Power in the Long Twentieth Century," *Diplomatic History* 33 (2009), 775–806.

own profile within policy-relevant knowledge networks which centered on the United States.

The Fulbright Agreement

On 17 May 1949 the governments of the United States and the Netherlands signed an accord to facilitate educational exchange between the two nations; the treaty went into effect exactly one month later. This bilateral Fulbright agreement was based on the provisions set out in the Surplus Property Act of 1944 and the subsequent Fulbright Act of August 1946, that allowed for the use of proceeds from the sale of surplus military equipment for educational activities. The two governments agreed to promote "a wider exchange of knowledge and professional talents through educational contacts" by "financing studies, research, instruction, and other educational activities" of US citizens in the Netherlands, and "furnishing transportation for nationals of the Netherlands [...] who desire to attend United States schools and institutions of higher learning." The agreement would be administered by a newly-created United States Educational Foundation (USEF), led by a board consisting of ten members, five Dutch and five American (of whom at least three had to be Foreign Service personnel).[13] The Foundation's initial address was at Bezuidenhout 18 in The Hague—a property since demolished to accommodate the Ministry of Economic Affairs.

In this environment, one might have expected that the addition of the Fulbright Program as an extra means to engage with US academia—both in terms of Dutch scholars residing in the United States and US professors coming the other way—would have been welcomed. But the Program did not emerge easily. The signing had been delayed due to Dutch complaints over the lack of equal provisions, and also due to American complaints over the low level of the Dutch financial contribution. It was finally agreed that the Netherlands would pay to the US Treasury a sum of Dutch Guilders equivalent to $250,000 annually, up to the point that this reached $5m (i.e., over the course of twenty years).[14] Although the original draft, dated 14 January 1948, had stated that the

---

13 "Fulbright Agreement between the United States and The Netherlands, June 17, 1949," Archive of the Fulbright Commission, Roosevelt Study Center, Middelburg (hereafter Fulbright archive), box G3: Research into USEF/NACEE History.

14 According to the Mutual Aid Settlement between the two nations, signed on 28 May 1947, "the two Governments agree that their rights and obligations in connection with the line of credit for the purchase of surplus property heretofore granted by the United States

Netherlands would only deposit the equivalent of $62,500 for the same calendar year, this was rejected by the Americans. Subsequent revisions of the Agreement in 1965 and 1972 did eventually see the Dutch government increase its stake to 50 percent of the costs, turning the Fulbright Program into a genuinely bilateral project.[15]

Similar complaints were expressed concerning the make-up of the USEF board. The original American plan proposed a board consisting of only seven members, with the US Chief of Mission at the embassy having "the power of appointment and removal of members of the Board at his discretion." Members from the embassy personnel would serve as chair and treasurer, and there would be only two Dutch members on the board. For an agreement intended to further mutual understanding, such an uneven division of power was deemed unacceptable by the Dutch government.[16] Similar proposals involving a 5:2 ratio of board members to the advantage of the United States had been resisted by other countries, notably Australia. France, soon followed by Italy, insisted on a 'commission' rather than a 'foundation,' and petitioned successfully for equal membership.[17] The Dutch government followed suit.

Finally, the uneven provisions for Dutch and American participants in the Fulbright Program also caused concern. While the Americans would receive full travel and accommodation allowances, the Dutch would only benefit from travel bursaries, thus needing to find other means to cover the costs of their stay in the US. While some historians—notably Jan Rupp—have claimed that this uneven arrangement represented some form of American insincerity, it was, in fact, caused purely by the lack of convertibility of the Dutch Guilder, and the USEF did not have a mandate to overcome this. The first annual report of the USEF in early 1950 stated the following:

Government in the amount of $30,000,000." The funds allocated for educational exchange therefore amounted to one-sixth of the Dutch debt accumulated through the purchase of surplus military equipment. Mutual Aid Settlement: Agreement and Exchange of Notes between the United States of America and the Netherlands, p. 7, Fulbright archive, box G3: Research into USEF/NACEE History.

15 "Agreement between the Government of the United States of America and the Government of the Kingdom of the Netherlands for the Financing of certain Educational Exchange Programs," [1972], Fulbright archive, box G3: Research into USEF/NACEE History. This also changed the name of USEF into Netherlands America Commission on Educational Exchange (NACEE).

16 "Draft of an Agreement," 14 January 1948, Fulbright archive, box G3: Research into USEF/NACEE History.

17 Frank Ninkovich, *The Diplomacy of Ideas: US Foreign Policy and Cultural Relations 1938–1950* (Cambridge, 1981), p. 141; Garner and Kirby, "Never a Machine for Propaganda?" p. 121.

> It is not thought that the Program has been received with particular enthusiasm in the Netherlands [...]. The negotiations dragged on and on and were finally concluded at a date which left little possibility of a successful first year program. Benefits to the Netherlands are not too apparent [...][.] Due to the inconvertibility of Dutch currency, it is the lack of dollars rather than a lack of guilders that controls the flow of travellers in all categories. Prior to the announcement of the details of the Program, there had somehow arisen a popular feeling that here was a new way of getting to America. Thus the actual possibilities came as a disappointment.[18]

The largest part of the budget in these years was allocated to supporting American lecturers, researchers, teachers, and graduate students to travel and stay in the Netherlands.[19] The popular image of the Program is that it provided a means for non-Americans to benefit from study in the United States. While this was certainly the case, its early provisions were more fully geared towards sending American specialists around the world so that they might spread their knowledge, values, and techniques. Those wishing to study and work in the US needed to secure one of the many university scholarships available for international students, which also pushed the main costs for running the Program in the United States on to the private sector.[20]

Numerically, however, it was Dutch graduate students in particular who would profit from the scheme: from 1949–1986, of the total number of 1999 Dutch Fulbright scholars, 1175 were graduate students.[21] In its early years the Program's selection procedure involved contacting professors with requests for recommendations of their brightest students for Fulbright scholarships.[22] Of the Dutch Fulbrighters between 1949–1997, 151 went on to become university

18 The agreement was signed on 17 May 1949, but the first award was not granted until October of the same year, and the Fulbright Commission did not have its own office space until November. Annual Report, USEF/The Netherlands, 15 February 1950, Fulbright archive, box A2: Annual Reports 1949–1963.

19 The estimated budget for 1954 was $338,000, of which $252,630 was reserved for covering the costs of the American participants. Annual Program Proposal: Program Year 1954, USEF/NL, Fulbright archive, box A1: Program Plans 1952–1969.

20 Maintenance grants for Dutch participants were finally introduced in 1974.

21 The American grantees, totaling 1396 for the same period, were more evenly spread among scholars, teachers, and students. "Overview of the Fulbright Program in the Netherlands since 1949," Fulbright archive, box G3: Research into USEF/NACEE History.

22 J.P. Middelburg, "Geschiedenis Fulbrightprogramma en NACEE," October 1997, unpublished paper, Fulbright archive, box G3: Research into USEF/NACEE History.

professors. This was essentially an enterprise based on academic merit. Sam Lebovic argues convincingly that the Program may have been pitched as furthering world peace through mutual understanding (or, rather, foreign scholars understanding the United States), but in practice it was "a means to influence international elites and lay the groundwork for broader political transformations," assuming that the brightest students selected for the Program would later form part of the socio-political and economic elites of their respective countries.[23] Efforts were made to maintain links with the participants over the longer term by means of a "follow-up program", albeit a discreet program that was carefully managed:

> Follow-up involves all the activities which are carried out in the United States and at foreign posts by which grantees and other visitors are encouraged to maintain contact with the United States after their return home [...]. It is Department and Agency (USIA) policy that returned exchange visitors will not be used in any way which will impair the prestige and credibility of the educational exchange program [...]. Returned exchange visitors must never be forced into situations where they are made to feel they must repay a debt through statements and activities pleasing to the United States. Neither should they be used in situations in which the public would gain the impression they are being exploited.[24]

The Dutch alumni association De Halve Maen was established as early as 1951, as a means to keep the Fulbright community together.

Despite the Program's goals to promulgate American know-how and techniques abroad, it was not easy to find suitable candidates to attend Dutch universities in the early years. The USEF Program Proposal for 1952–1953 commented that "very few of the vacancies for United States Visiting Lecturers [...] were actually filled." This was in part due to the fact that local universities tried to use the Program to obtain the services of US experts who were too much in demand—a pattern that recurred in many nations with Fulbright agreements in the Program's early years. When scholars proved to be unavailable (or unwilling) to travel to the Netherlands, it was not possible to find alternative candidates to replace them. In the early years, American participants were selected to attend Dutch universities according to three factors: the need of

23 Lebovic, "From War Junk to Educational Exchange," p. 306.

24 "Maintaining Contact with Returned Grantees," Bureau of Educational and Cultural Affairs, Department of State, 11 October 1961, Fulbright archive, box G3: Research into USEF/NACEE History.

the host institution; the possibility for "the introduction of current American conceptions, methods, and techniques;" and finally US scholars needed to be distributed across Dutch institutions such that no institution would have more than one during any academic year.[25] Nevertheless, in 1955 Amsterdam professor and Fulbright participant Arie den Hollander summed up the thoughts of many:

> The Fulbright Act has made a unique mobilization of forces possible in training, the arts and science. Incidental criticisms of how the Fulbright Act is implemented are heard, but this does not undermine the scale of the vision, the generosity of the decision and the favorable influence of the possibilities for tens of thousands to encounter research techniques and educational methods elsewhere, the lifestyles of other peoples, and the removal or correction of many of the existing stereotypes and other dangerous prejudices.[26]

Given the respective resources and knowledge bases in the United States and the Netherlands, the Fulbright Program was an obviously asymmetrical enterprise: the US impact in the Netherlands would be far greater than the other way around. Yet it would be simplistic to see this purely as a one-sided encounter. The Program enabled Dutch institutes to enter into US-based knowledge networks, to spread their technical know-how, and to raise their profile as a result. Institutions such as the Technical College in Delft were able to benefit in fields such as hydrodynamics, aerial mapping, and construction in this way.[27]

After 1945, the Dutch were keen to revive their scholarly contacts with the United States. The Rockefeller Foundation had provided grants for Dutch scholars throughout the 1920s and 1930s. Arie den Hollander was a beneficiary of a Rockefeller grant in 1932–33 which he used for fieldwork for his *De landelijke arme blanken in het Zuiden der Vereenigde Staten: Een sociaal-historische en sociografische studie* (1933). In 1946 den Hollander was made professor of sociology in Amsterdam, and a year later he added "Amerikanistiek"—American Studies—to his portfolio. Following a stint as a Fulbright research scholar at the University of Michigan in 1951–1952, he founded the *Amerika*

25 United States Educational Foundation in the Netherlands (hereafter USEF/NL), Program Year 1952, Fulbright archive, box A1: Program Plans 1952–1969.

26 A.N.J. den Hollander, "James William Fulbright," 17 November 1955, Fulbright archive, box G3: Research into USEF/NACEE History.

27 A.J.M. de Leeuw, "Het Fulbright programma in Nederland," *TH Mededelingen* 10 (1962), 2–3. Delft became a Technical University in 1986.

Instituut at the University of Amsterdam in 1952; this was the first step towards establishing American Studies institutionally within Dutch higher education.[28] The Harvard-Leiden Foundation was set up in 1945 to assist with an annual exchange of professors and students between the two prestigious universities. Several prominent figures, including Arthur Schlesinger Jr. (1948–1949) and Harvard history professor Perry Miller, came to study and teach in Leiden for one academic year each.[29]

More significantly, the Netherlands-America Institute (NAI) was created in 1946 by a group of Amsterdam businessmen. Its origins were both symbolic—to thank the United States for its sacrifices in World War II—and practical—to establish opportunities for apprenticeships in the United States. In 1948 the NAI merged with the New York-based Netherlands-America Foundation, a private body that had been founded in 1923 for the same purpose, but which had been moribund since the 1930s, when the Depression affected its operations. By the time a Fulbright agreement was signed in 1949, the NAI had been actively selecting candidates for a variety of scholarships in the United States for four years. The Dutch government recognized the Institute as the sole representative for dealing with educational exchanges concerning Fulbright and the Information and Educational Exchange (Smith-Mundt) Act, and a subsidy was provided for the NAI to fulfill this role. In 1951 the Institute could claim that "practically all applications for American [educational] grants, both govern-

28 The Institute's library was developed with the help of donations from the Rockefeller Foundation, the Library of Congress, and the New York Public Library. Den Hollander visited the United States in 1946 and 1948 to gather support, and in 1950–51 spent another year there as a Fulbright lecturer. He would also play a key role in the foundation of the European Association of American Studies via the Salzburg Seminar in 1954. Archive of professor A.N.J. den Hollander, University of Amsterdam, Universiteitsmuseum, Amerika Instituut, folder 82, 83, 166; Alfons Lammers, "Hollander, Arie Nicolaas Jan den," *Biografisch Woordenboek van Nederland* 2 (1985), 238–40; Peter Jan Knegtmans, *Een kwetsbaar centrum van de geest: de Universiteit van Amsterdam tussen 1935 en 1950* (Amsterdam, 1998), pp. 276–8.

29 Unfortunately Miller would write a damning criticism of European universities in the *Atlantic* magazine in 1951, which did not help to prepare the ground for Fulbright. See Perry Miller, "What drove me crazy in Europe," *The Atlantic*, March 1951. Tity de Vries, *De Amerikaanse cultuurpolitiek ten aanzien van Nederland 1945–1960* (MA thesis, University of Groningen, 1983), pp. 46–8; Nicolas Guyatt, "An Instrument of National Policy: Perry Miller and the Cold War," *Journal of American Studies* 36 (2002), 107–49, on p. 132. The Harvard-Leiden program was transformed in 1949 into the Nederlandse-Amerikaanse Universitaire Stichting, a foundation with a broader mandate to encourage inter-institutional exchanges in general.

mental and non-governmental, are submitted to the Screening Committee."[30] In the years up to its closure in 1976 the Institute would broaden its function as a culture and information center and library, playing an important intermediary role between Dutch society and American culture.

Lastly, there was the *Nederlandsche Opleidings Instituut voor den Buitenlandschen Dienst* (NOIB: Netherlands Training Institute for Foreign Service). Situated at Nijenrode outside of Utrecht, it was founded in 1946 as the first institute in the Netherlands to offer management training along the lines of the American business schools. To ensure political independence, funds were provided by the major Dutch multinational corporations (KLM, Unilever, Philips, and Shell) who were interested in a state-of-the-art training program that could function as a source of skilled management personnel for their worldwide operations. From the very beginning the American business school system was adopted as the model for the curriculum, and from 1950 onwards Nijenrode not only received a Fulbright lecturer every year but also ran annual student exchanges with US business schools.

Alongside the Dutch goal to integrate Dutch expertise into US knowledge networks, the Fulbright Program also enabled the United States to draw on European expertise to further its own technological advantages. It was weighted towards American interests by promoting American ways of thinking and doing amongst its allies, but allies could also benefit in the process. In this respect the Fulbright Program was one of many initiatives in this area; others included the Foreign Leader and Specialist Programs, the American Council of Learned Societies scholarships, the Ford Foundation Travel and Study Grants, the prestigious Harkness scholarships of the Commonwealth Fund, and the Salzburg Seminars. Collectively these could be spread out across different sectors of society or, should the need be seen, directed towards a specific institution or field of study of interest to US public diplomacy at any particular time, which would intensify the impact. The Fulbright Program is, therefore, only one part of a wider story; but its two-way exchange sets it apart as a unique enterprise in 'knowledge transfer.'[31] The American Fulbrighters—

30 NAI-NAF Annual Report 1951, Archive Ministry of Foreign Affairs, The Hague (hereafter BZ), box 33: Netherlands-America Foundation 1945–1954, 810.2, Code 8 1945–1954.

31 See Henk te Velde, "Political Transfer: An Introduction," *European Review of History* 12 (2005), 205–21; D.P. Dolowitz and D. Marsh, "Learning from Abroad: The Role of Policy Transfer in Contemporary Policy Making," *Governance* 13 (2000), 5–24; Russell Prince, "Policy Transfer, Consultants and the Geographies of Governance," *Progress in Human Geography* 36 (2012), 188–203; Michael Espagne and Michael Werner (eds.), *Transferts. Les relations interculturelles dans l'espace franco-allemand* (Paris, 1988); Johannes Paulmann,

often established in their field, and holding great prestige—regularly functioned as catalysts of change and proselytizers of American teaching and research methods abroad. They lectured in universities, to professional organizations, and to the public; they established and ran research groups; they published, sometimes with Dutch co-authors, in Dutch and international journals; they wrote on Dutch subjects or the state of the field in the Netherlands; they acted as consultants on educational or professional matters; they encouraged Dutch graduate students to attend US universities; they maintained their contacts with the Dutch institutions after they returned to the US. A pattern emerges: it is possible to trace how US academic 'pioneers' (particularly in the social sciences) came to the Netherlands in the 1950s and 1960s, contributing to the popularity and development of these disciplines at Dutch universities. By the 1970s and 1980s Dutch academics (and particularly graduate students) in these (new) disciplines traveled to the US for further study.[32] The Fulbright Program was in many ways a classic exercise of hegemonic power, whereby the US hegemon provided opportunities for allies to 'integrate' themselves into its modes of behavior, in the process 'socializing' behavior around certain norms and creating pay-offs for both; this is what John Krige refers to as the 'co-production' of hegemony.[33]

The Program was flexible in that different subject areas could be promoted according to the needs of a particular period. In 1950, the first full year of the Program, 70 percent of Dutch Fulbright scholars sought out opportunities in the sciences and medicine. For the Dutch it was simple: they would benefit from American know-how as a means to a good career, and there were more generous scholarships from US universities available in the sciences than in the humanities. For USEF (and NAI) the primary goal was to spread a broader understanding of American values and culture—broadly speaking, it was meant as a cultural as much as a learning experience. This fit with the promotion of American Studies in the Netherlands, which became a stronger focus in the late 1950s and early 1960s. Due to the fact that American Studies did not exist at Dutch universities prior to the foundation of the Amerika Instituut in Amsterdam in 1952, it took some time before this had an effect. The multi-

"Internationaler Vergleich und interkultureller Transfer: Zwei Forschungsansätze zur europäischen Geschichte des 18. bis 20. Jahrhunderts," *Historische Zeitschrift* 267 (1998), 649–85; Everett Rogers, *Diffusion of Innovations* (New York, 1962).

32 Rupp, "The Fulbright Program," table on p. 80.

33 See John Krige, *American Hegemony and the Postwar Reconstruction of Science in Europe* (Cambridge, MA, 2006); John Krige, "Building the Arsenal of Knowledge," *Centaurus* 52 (2010), 280–96.

disciplinary approach of American Studies, with its cross-over identity between History, Cultural Studies, and the Social Sciences, also contributed to its slow implementation in the formal departmental orderliness of Dutch higher education institutions. The Fulbright Program, by linking up with other forms of philanthropic or state patronage, could provide a valuable stimulus at particular junctures.[34]

The USEF Annual Report for 1960 remarked off-handedly that "[o]ver the years a certain pattern of projects has evolved consisting partly of fields in which the Program can be of assistance to Netherlands institutions of higher learning and partly of subjects which the Foundation feels should be encouraged."[35] The wording is revealing—on both counts the USEF saw itself as the key player determining what should and should not be promoted. The Program displays clear trends—for example, the rise of interest in US social and political sciences in the 1960s is quite obvious—but the overall coverage remained broad. For 1949–1990 the most popular discipline for Dutch Fulbrighters was law, closely followed by medicine, but both of these amounted to no more than 6 percent of the total number of scholars respectively. Psychology and History made up 5 percent each (for the distribution of grants among fields during 1949–1960, see Figure 7.1). A 1975 NACEE survey of the Program listed the academic fields in which "positive results" had been achieved: American Studies; Education; Urban and Environmental Studies; Linguistics; Political Science; Computer Science; Pastoral Counseling; Civil Engineering; Demography.[36]

From 1953 onwards, detailed reporting under 'Evidence of Effectiveness' became a standard feature of USEF Annual Reports. The lists include lectures and public engagements given by US Fulbright professors, and the articles that they published in Dutch publications (sometimes with Dutch co-authors). Plenty of examples are available of US Fulbrighters contributing to both the national and the European research climate. In 1952 research scholar George E. Boyd contributed as a visitor to Leiden University in the field of Physical Chemistry, delivered ten individual lectures at other Dutch universities and co-chaired the International Congress on Pure and Applied Chemistry in Stockholm. In the subsequent year, research scholar Joe M. Smith was placed at Delft Technical College, and delivered a lecture entitled 'The Future of Chemical Engineering Education in the United States' at the Royal Institute

34 See Scott-Smith, "The Ties That Bind."

35 USEF Annual Report: Program Year 1960, 13 September 1960, Fulbright archive, box A2: Annual Reports 1949–1963.

36 "Major Achievements since 1970—Netherlands Program," Fulbright archive, box G3: Research into USEF/NACEE History.

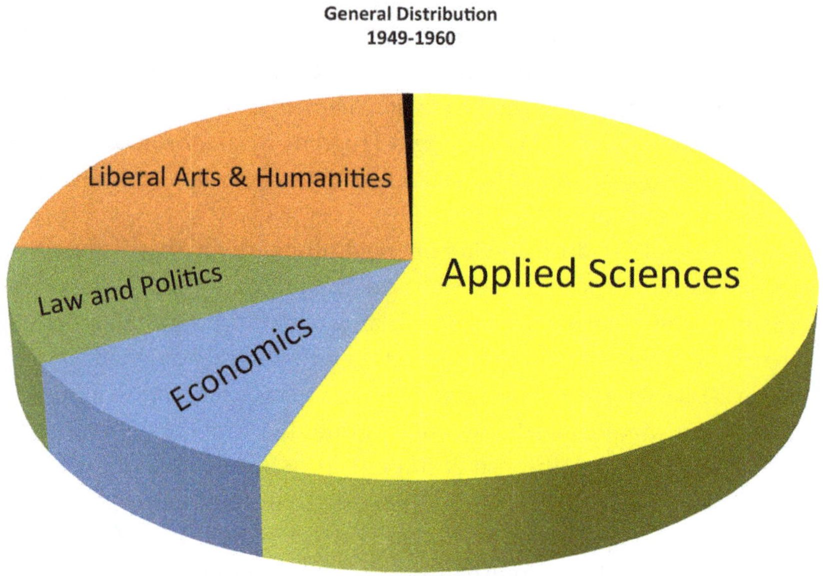

FIGURE 7.1 *Distribution of Fulbright grantees among different disciplines in the years 1949–1960, based on data provided by the Netherlands Fulbright office. 'Applied Sciences' covers all the exact and life sciences, and engineering. For some Fulbright scholars no subject is listed, and this group is represented by the small black segment of the pie chart.*

of Engineers. In the same year, research scholar John W. Oswald conducted research on crop diseases at Wageningen University and discovered two new variants. 1956 saw research scholar Robert E. Bolinger lead a research group at Wilhelmina hospital in Amsterdam; results of their research on insulin were published in the *Nederlands Tijdschrift voor Geneeskunde*. Other members of his team used results of the same research as a basis for lectures at other Dutch institutions. In 1957 a visiting lecturer in Psychology at the University of Amsterdam, Professor Frederick B. Davis, greatly influenced educational science and psychology in the Netherlands. Staff at the Institute for Pedagogics and Didactics in Amsterdam, under the leadership of Professor A.D. de Groot, constructed reading tests for use in Dutch schools that made use of Davis's materials. Davis continued to act as an academic advisor to the Institute after his return to the United States.[37]

37 See also John C. Flanagan, "Frederick Barton Davis 1909–1975," *Psychometrika* 41 (1976), 4–7.

Other results from that year involved Dr. Ronald Freedman, an Assistant Professor of Psychology at the University of Michigan who was also a visiting lecturer in Amsterdam, and delivered the tenth-anniversary speech at the Institute of Social Research in Utrecht; he also acted as a consultant for the Central Bureau of Statistics; Dr. Herbert Gold, research scholar in Plant Pathology visiting Wageningen, intended "to encourage the exchange of scientists between the laboratory for plant pathology of the University of California and Dutch laboratories. With the support of the head of his department at Berkeley, Dr. Gold plans to request funds to support a Dutch research worker and his family at the American laboratory for a year. In these plans Dr. Gold will be continuing efforts started by Dr. J.W. Oswald of the same laboratory, who was a 1953/54 Fulbright grantee in the Netherlands."[38] 1958 saw Dr. Hirsch Cohen, research scholar in Mathematics at Delft, be a co-author of several papers on cavity flows in collaboration with the Technical College's Jan Geurts. Cohen further spread his Delft research results via lectures in Denmark and Israel. 1960 brought Professor Homer B. Metzger to lecture on marketing in Wageningen, where "[t]his subject was hitherto not taught [...]. Responsible university officials are planning to institute such a course, however, and the work done by Professor Metzger is considered to have prepared the ground."[39] That same year saw Dr. Joseph Kuc, assistant professor in Biochemistry at Purdue College, research "biochemical factors responsible for disease resistance in plants at the Phytopathological laboratory at Wageningen [...]. Professor Kuc was invited to remain at the Agricultural College, Wageningen, as a permanent staff member with the rank of professor extra-ordinary."[40]

While these US embassy reports are obviously aiming at the positive promotion of the results of educational exchange, a noteworthy element to all these narratives is the mobility granted to US scholars by the Fulbright apparatus. The Inter-Foundational Lecture Program, operated by US embassies, enabled American Fulbright scholars to tour Western Europe in order to deliver lectures, thus maximizing their audience: in May 1955 Morris Janowitz, a University of Michigan sociologist based at the University of Frankfurt as a research scholar, lectured in Amsterdam on 'Political Behaviour;' in April 1957 Dr. W.W. Boone, a

38 USEF Annual Report: Program Year 1957, 13 September 1958, Fulbright archive, box A2: Annual Reports 1949–1963.

39 USEF Annual Report: Program Year 1960, 13 September 1960, Fulbright archive, box A2: Annual Reports 1949–1963.

40 Idem.

visiting lecturer at the Blindern Mathematical Institute in Norway, presented at the Institute of Fundamental Research in Amsterdam on 'An attempt to outline the full proof of the insolvability of the world problem' (the text of the lecture is unfortunately lost). Moreover, American Fulbrighters often wrote on Dutch subjects for an English-language audience, once more integrating Dutch knowledge into US-centered academia. For instance, Theodore M. Brown published *The Work of Gerrit Rietveld, Architect* (1958); H.H. Feestehausen produced *Economic Knowledge and Comprehension in a Netherlands Farming Community* (1965); John P. Windmuller covered *Labor Relations in the Netherlands* (1969).

Cold War Science

Science was naturally at the center of US interests after WWII. It represented the future, the superiority of the democratic system that supported it, and, above all, power. The stature and quality of American science proved attractive to foreign scientists. Through its fellowships program, the Rockefeller Foundation had initiated a long-term strategy to create networks of scientific training and expertise circulating through the United States before WWII. After the war this process grew to be more state-driven and militarized, as the Cold War contest with the Soviet Union turned the political allegiance of scientific expertise into a national security issue (something that already began with the Manhattan Project). Yet, after the 'Red Scare' of Truman's second term, the democratic aspect was finally emphasized in President Eisenhower's Atoms for Peace speech to the UN in December 1953. It stated the intention to promote the civilian use of nuclear power and—crucially—to provide access to nuclear technology for other nations. The speech projected an image of the US as the leading peaceful nation. The Ford Foundation joined this process in 1956 by providing grants for Niels Bohr's Institute in Copenhagen, specifically to promote the circulation of scientists, skills, and ideas in Europe and beyond.[41]

Dutch science was badly affected by WWII, both in terms of loss of materiel and of loss of Jewish personnel. There was every reason to want to seize opportunities for re-connecting with the outside world, and the United States, with its funding and research base, was the most logical location to achieve the most rapid results for rebuilding science. The nuclear explosions of August 1945 were the direct catalyst behind the formation of the Foundation for Fundamental Research on Matter (FOM) in 1946. It was intended to reinvigorate the Dutch scientific infrastructure and expertise in fundamental physics. The initial discussions of this development already highlighted the need

41 Krige, *American Hegemony*, p. 180.

for Dutch scientists to travel to the United States as an essential part of their training. Without this re-connection to the top levels of international physics research, the Netherlands might be excluded from major discussions at the diplomatic level. Therefore, "renewing and maintaining contact with Americans already figured prominently in Dutch scientists' earliest plans and actions."[42] The FOM, together with the broader Dutch Organisation for Pure Scientific Research (ZWO), had a major impact on the reconstruction of scientific structures. FOM's budget was originally planned to rise to 5m guilders, but even the much smaller sum of 0.5m guilders which was actually provided in 1947 made for a substantial contribution in a period of post-war austerity. The intent of large-scale funding was also a way to gain the attention of the Americans, and investments in nuclear physics, mathematics, and aerodynamics heightened the interest in transatlantic connections.[43]

The Dutch government was, therefore, keen to promote the circulation of Dutch scientists in the United States in the post-war years. For instance, physicist Abraham Pais took up a fellowship with Niels Bohr in 1946, and went on to the Institute for Advanced Study at Princeton, for research into particle physics. FOM did not have much travel funding readily available, but was able to support certain projects successfully. Thus physicist Siegfried Wouthuysen was able to study with Robert Oppenheimer in 1947 before returning to his research post in Amsterdam. Physics had a strong tradition in the Netherlands dating back to the turn of the century which had featured key figures such as Hendrik A. Lorentz, Johannes van der Waals and Heike Kamerlingh Onnes. However, experimental nuclear physics had not developed sufficiently, resulting in a lack of substantial knowledge in this field. Input from private industry was deemed vital to boost research and maintain the link with commercial possibilities. It was this direction that led to the formation of the Institute for Nuclear Physics Research (IKO) as a subsidiary of the FOM in 1946, also supported by the city of Amsterdam and Philips.

The Fulbright Program arrived at an ideal time for promoting further exchanges between Dutch and American scientific communities. Eisenhower's Atoms for Peace initiative took this development even further. Together with the FOM, the Dutch government prepared a bilateral agreement that was discussed in Washington DC on 28 February 1955. The aim was to include the exchange of classified information at a point in the future. With this prospect in mind, a test

42 Friso Hoeneveld and Jeroen van Dongen, "Out of a Clear Blue Sky? FOM, The Bomb, and The Boost in Dutch Physics Funding after World War II," *Centaurus* 55 (2013), 264–93, on p. 274.

43 Rupp, *Van oude en nieuwe universiteiten*, pp. 214–6.

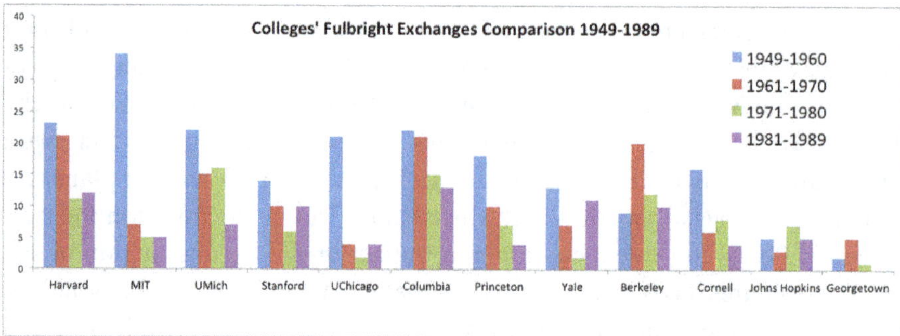

FIGURE 7.2 *Destination of Dutch Fulbright scholars for various periods.*

reactor was proposed at Petten, and the Technical College in Delft launched a new course on nuclear reactor technology. Together, these steps created a new nuclear research cluster that involved Delft, FOM, Petten, and expertise at other Dutch universities.[44] During the 1950s close working relations were developed between the FOM, the Norwegians, and then the British, around nuclear technology and uranium supply.[45] The driving force behind these moves was Dutch theoretical physicist Hendrik A. Kramers, a former close collaborator of Niels Bohr, who had previously established high-level contacts in the United States. Inspired by Vannevar Bush's *Science, The Endless Frontier* report of 1945, Kramers was close to MIT president James Killian, Eisenhower's special assistant for science and technology. He was also a diplomat, serving on the Scientific and Technological Committee of the UN's Atomic Energy Commission.[46] It is both clear and noteworthy that MIT became a key destination for Dutch research scholars in the hard sciences following the Atoms for Peace initiative in the mid-1950s (see figure 7.2).

In physics, several Dutch 'pioneers' made use of the Fulbright Program for access to the latest developments in American research—and to make use of American facilities to pursue their own explorative experimental work. Alex Verrijn Stuart was one of the first Dutch pioneers: in 1950, as a graduate student, he was sent to the University of Michigan, where he obtained his PhD. He was eventually appointed the first professor in computer science in Leiden

44 M. Wessels, *De kerncentrales van Dodewaard en Borssele* (Den Haag, n.d.).

45 J.M. van Splunter, "Love at First Sight: Co-operation Between the Netherlands and Norway on the Peaceful Use of Atomic Energy, 1950–1960," *Institutt for forsvarsstudier* 2 (1994), 1–47. See also his "Nuclear Fission across the North Sea: Anglo-Dutch Cooperation on the Peaceful Use of Atomic Energy, 1950–63," *Journal of Contemporary History* 29 (1994), 663–709.

46 Hoeneveld and van Dongen, "Out of a Clear Blue Sky?", on p. 278.

in 1969, and in 1976 he was one of the founding members of the IFIP TC8 Technical Committee of Information Systems. Hendrik den Hartog is another example of a Dutch pioneer. He became a research scholar at the University of Pennsylvania in 1951, and ended up as Professor of Biophysics in Amsterdam (1962). J.C. Kluyver, co-founder of the biophysics institute in Utrecht with Rockefeller Foundation support, moved to Delft after WWII and was a research scholar at MIT in 1951. He later became Professor of Physics in Amsterdam. Leiden researcher Jakob Kistemaker had studied at Niels Bohr's institute in Copenhagen before being appointed head researcher for the FOM's Laboratory for Mass Spectography in Amsterdam in 1949. Kistemaker achieved notoriety when, as a Fulbright research scholar in Chicago, he "surprised the Americans with the production of enriched uranium (in which the Americans wished to have a monopoly)" through the ultracentrifuge process.[47] To what extent this anecdote is based on true events remains unclear, but there is no doubt that Kistemaker's expertise raised the awareness of Dutch scientific achievement in the United States. He would be appointed the head of the Institute for Atomic and Molecular Physics (AMOLF) in 1955. Other major figures on the Fulbright participant list include Nico M. Hugenholtz, a research scholar at Princeton in 1958, who later became Professor of Physics in Groningen; Jan Hilgevoord, a research scholar at Princeton in 1960, who became Professor of Theoretical Physics at the University of Amsterdam in 1967; Leonard Rietjens, who arrived at the University of Michigan as a research scholar in nuclear physics in 1957, the same year that he published the report *Kernphysica en kernenergie* for the Dutch Royal Navy Institute. Rietjens would later be the notorious promulgator of magneto-hydrodynamic energy conversion, whereby electrical energy is released through heating gasses to extreme temperatures, and then 'broken up' by magnetic fields. American and Russian interest in the technique prompted the Dutch Minister of Economic Affairs, Ruud Lubbers, to support a prototype; but the debilitating costs of developing a commercial version caused the end of the project in 1986. Nicolaas Bloembergen was already familiar with the American scientific environment when he became a Fulbright research scholar at Harvard in 1950. After departing for the United States for graduate studies in physics at Harvard in 1945 he was involved with Edward Purcell's experimentation in nuclear magnetic resonance at MIT's Radiation Laboratory. Bloembergen, who returned to the Netherlands due to the new FOM funding,

47 Klaas van Berkel, Albert van Helden and Lodewijk Palm (eds.), *A History of Science in The Netherlands. Survey, Themes and Reference* (Leiden and Boston, MA, 1999), pp. 230–31; Hoeneveld and van Dongen, "Out of a Clear Blue Sky?", on p. 283. See also the contribution by Abel Streefland in this volume.

submitted his PhD on Nuclear Magnetic Resonance to Leiden in 1948, and after a brief spell working with his supervisor Cor Gorter he was able to use the Fulbright award to return to the United States, where he became an associate professor, at Harvard, in 1951.

The connection between Dutch nuclear physics and US-based research was also aided by the arrival of American experts in the Netherlands via the Fulbright Program. In 1954–1955 the Manhattan Project scientist Professor John R. Huizenga, then active at the Argonne National Laboratory, spent a year in Amsterdam as a Fulbright research scholar. At the time Huizenga was renowned for his contribution to the discovery of two new elements, einsteinium and fermium; their discovery was a result of the first thermonuclear explosion in 1952.[48] N.M. Hugenholtz has also noted the influence of John Gammell, a Los Alamos scientist, on nuclear science in Holland; Gammell was a visiting researcher at Utrecht University in the mid-1950s. Hugenholtz spent 1952–1954 at Niels Bohr's institute in Copenhagen, before taking a two-year research position at Princeton. Such an arrangement was "fairly ordinary" for Dutch scientists at that time, and a stint in the United States was regarded an important element of career advancement. After gaining his PhD in Utrecht in 1958, Hugenholtz's US visit was fully funded by the FOM and was extended from one to two years without a problem. Further trips to Iowa, Madison, and the University of Pennsylvania followed.[49] For an overview of the names of physicists who received Fulbright grants, see Table 7.1 in the Appendix to this article.

It is striking that research scholars rather than students dominate the list of physicists. This indicates that in the first decade of its existence, the Fulbright program was predominantly a means for established scholars to experience the top-level research environment of the United States. This was the most effective way to engage the Netherlands with US-led research programs after World War II. As mentioned above, also noteworthy is the dominant position of MIT on the list of destinations; this may indicate the influence of Bush and Killian in steering Fulbright scholars to that institution, along with the abundance of research funding available from government contracts there (see also Figure 7.3).

A similar development can be observed in the field of engineering (see Table 7.2 in the Appendix); here, Dutch experts in electrical, aeronautical, and hydrological engineering utilized the Fulbright funding to explore research opportunities in the US. Hans van Nauta Lemke is a prime example. Having

48 See John R. Huizenga, *Five Decades of Research in Nuclear Science* (Rochester, NY, 2010).

49 N.M. Hugenholtz, telephone interview, 13 December 2013.

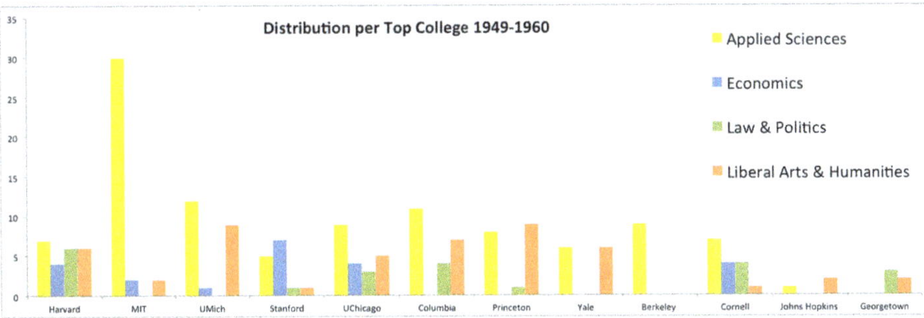

FIGURE 7.3 *Distribution of Dutch Fulbright grantees to top-level US universities according to subject area, 1949–1960.*

completed his studies in electro-technical engineering in Delft in 1950, van Nauta Lemke was employed by the appliance manufacturer Van der Heem, and soon became one of their chief innovators, particularly on military contracts. Early projects for the Dutch Navy included the improvement of television antennae and the development of a submarine detection system for use in torpedo boats. In 1954 van Nauta Lemke submitted a research proposal to the US and won a scholarship, even though as a professional (rather than academic) he was not strictly speaking eligible for a Fulbright award. He spent several months of the following year at MIT, studying transistor technology and combined this research with free-lance work for Van der Heem at the weekends. The Dutch company had not initially agreed to his trip, but were happy with an agreement for him to continue work for the company while abroad. Van Nauta Lemke was also to examine the production of silicon transistors at Texas Instruments. On his return van Nauta Lemke was named head of Van der Heem's electrical apparatus laboratory, and in 1959 he became professor in electrical engineering in Delft. He has referred to the Fulbright trip as vital for his career; however, the circumstances of his unusual access to the program, as a non-academic, remain unclear.[50] Van Nauta Lemke continued working on contracts for the military throughout the 1960s, and it may have been the case that his expertise in dual-use technologies resulted in the Dutch state putting him forward as a preferred candidate. In any case, MIT was also the preferred destination for engineers, but there were far more graduate students among them than among the physicists. This may indicate the sense of an even greater urgency to upgrade Dutch engineering at the time, perhaps because of a lack of qualified teaching personnel post-WWII, and thus a struggle to expand the subject area in the Netherlands.

50 Hans van Nauta Lemke, telephone interview, 16 December 2013.

Conclusion

There is no doubt that the Fulbright agreement signed between the Netherlands and the United States in 1949 greatly increased the opportunities for scholarly exchange between the two countries. What is most interesting about this arrangement is the perspective of the Dutch scientific authorities on the Fulbright program: determined to revive the level of research and the mobility of its personnel after World War II, they considered the Fulbright a fast-track access to the best higher education and research establishments that the United States had to offer. The delay in finalizing the agreement was entirely based on the Dutch realization that the terms for educational exchange, as determined by the United States, were not as altruistic as had initially been assumed. Once this matter was settled, the FOM and ZWO made full use of this opportunity, and thereby aligned the Dutch scientific establishment closely with that of the leading nation, the United States. Nuclear physics was at the forefront of this alignment, but other areas, notably engineering, mathematics, chemistry, and medicine, also figure prominently in the Fulbright exchanges of the first decade. The Netherlands also boasted a significant research capacity (with dual-use possibilities) in these subjects.

Of course, the exchange of scientific expertise was only one facet of the Fulbright's mission. As the executive director of the Netherlands-America Committee for Educational Exchange during 1976–1992 commented in an interview, the numerous Dutch students who attended US universities thanks to the Fulbright often returned as "people of the world," and the first-hand experience of living and studying in the United States was "an eye-opener."[51] The list of well-known Dutch men and women who encountered the United States for the first time via a Fulbright exchange is impressive, ranging from politicians such as deputy prime minister Jan Terlouw (student of social work at Case Western Reserve University in 1958) and EU commissioner Frits Bolkestein (undergraduate at Smith College in 1962–1963), to mathematician, governmental advisor and public intellectual Alexander Rinnooy Kan (research scholar at Berkeley in 1976) and Nobel laureate and economist Jan Tinbergen (research scholar at Harvard in 1956). Particular subject fields, especially psychology, sociology, and political science, were heavily influenced by post-war trends in the United States, and several 'pioneer' Dutch academics made use of the 'Fulbright opportunity' to become acquainted with American

51 Joke Wind, interview with J.P. Middelburg, n.d. [1994–1995], Fulbright archive, box G3: Research into USEF/NACEE History.

methods and approaches, in order to instill them in their departments after their return home.[52]

The data provided by the Fulbright Program offers a unique insight into the transformation of the post-war Dutch academic landscape under the influence of the opportunities, techniques, and connections available in the United States. This transformation can be mapped out across the disciplines, according to the timing of the educational exchange, its geographical and cultural direction (Dutch scholars to the US, US scholars to the Netherlands), the resulting products of knowledge (lectures, publications, research teams), and the adaptation of Dutch educational culture to American norms and values over time. Particular studies have so far only highlighted the impact of the Fulbright Program on selected disciplines, whereas the possibility now exists for a full-spectrum assessment of US influence across Dutch academia over several decades. While other influences—primarily intra-European—were naturally also active (e.g. Niels Bohr and the CERN project in fundamental physics), the consistent level of US patronage, the commitment of Dutch authorities to utilizing the same, and the enduring attraction of US higher education led the Fulbright Program to exert a special influence on post-war Dutch academia.

Appendix: Tables

TABLE 7.1 *Dutch physicists who received Fulbright grants to conduct research in the United States, 1949–1960.*

Year	Scientist	NL University	US University	Position
1949	W.G. Burgers	Delft	Purdue	Research Scholar
1949	B.R. Nijboer	Utrecht	Princeton	Research Scholar
1950	N. Bloembergen	Leiden	Harvard	Research Scholar
1950	A. Verrijn Stuart		Michigan	Graduate Student
1950	H. de Vries	Groningen	Chicago	Research Scholar

52 See, for instance, Sjaak Koenis and Janneke Plantenga (eds.), *Amerika en de sociale wetenschappen in Nederland* (Amsterdam, 1986).

TABLE 7.1 *Dutch physicists who received Fulbright* (cont.)

Year	Scientist	NL University	US University	Position
1951	K. Eriks		Penn State	Research Scholar
1951	J.C. Kluyver	Delft	MIT	Research Scholar
1951	H. den Hartog	Amsterdam	Penn State	Research Scholar
1951	J. Kistemaker	Leiden	Chicago	Research Scholar
1951	F.L. Stumpers		MIT	Research Scholar
1953	P.H.E. Meijer		Case Institute	Research Scholar
1953	N. Poulis		MIT	Research Scholar
1954	R. Braams		MIT	Research Scholar
1954	C.H. Paris		MIT	Research Scholar
1955	G.W. Barendsen		Yale	Research Scholar
1955	J.J.M. Beenakker		MIT	Research Scholar
1956	P.S. Dubbeldam		MIT	Research Scholar
1956	G.J. Hooyman	Utrecht	Michigan	Research Scholar
1956	H.C. Kramers	Leiden	Ohio State	Research Scholar
1956	G.W. Tichelaar		Illinois	Research Scholar
1956	B.P.Th. Veltman	Delft	MIT	Research Scholar
1957	L.H.Th. Rietjens	Eindhoven	Michigan	Research Scholar
1958	N.M. Hugenholtz	Utrecht	Princeton	Research Scholar

Year	Scientist	NL University	US University	Position
1958	R. Wageningen	Groningen	MIT	Research Scholar
1958	C. Dullemond	Nijmegen	Rochester	Student*
1960	J. Hilgevoord	Utrecht	Princeton	Research Scholar

* Returned to Rochester as a Fulbright Research Scholar in 1979
Source: Fulbright Alumni 1949–1961, Dutch Fulbright Office (1990).
Note: some data about home institutions were unavailable, in which case the cell has been left blank.

TABLE 7.2 *Dutch electrical, aerodynamic, hydrological, and chemical engineers who received Fulbright grants to conduct research in the United States, 1949–1955.*

Year	Scientist	NL University	US University	Position
1949	C.J.M. Holstege	Delft	Oregon State	Graduate Student
1950	A. Delsman		Princeton	Graduate Student
1950	H.J. Hassig	Delft	MIT	Research Scholar
1950	J. de Heer	Delft	Ohio State	Research Scholar
1950	H. Krusemeyer		MIT	Lecturer
1950	S.M.L.van der Meer		Pennsylvania	Graduate Student
1951	H. Bosch		MIT	Graduate Student
1951	H. Coster		Purdue	Graduate Student
1951	J. Korringa		Ohio State	Lecturer
1951	N. Luning Prak		Columbia	
1951	R.J. Schliekelman		MIT	Research Scholar

TABLE 7.2 *Dutch electrical, aerodynamic, hydrological, and chemical engineers* (cont.)

Year	Scientist	NL University	US University	Position
1951	A.A. van Trier		Columbia	
1951	L. Troost		MIT	Research Scholar
1952	R.D. Bleeker		MIT	Research Scholar
1952	E.W. Groneveld	Twente	MIT	Research Scholar
1952	G. Haaijer		Lehigh	Graduate Student
1953	J.Th. Thijsse	Delft	Ann Arbor	Lecturer
1953	J.P. Verschuren		Lehigh	Graduate Student
1953	J.P. de Vries		MIT	Lecturer
1954	M.J. Brabers		MIT	Research Scholar
1954	C. Krijgsman		Stevens Institute of Technology	Graduate Student
1954	B. Okkerse		Illinois	Research Scholar
1954	M.P.C. van Overstraten Kruysse		Harvard	Graduate Student
1955	P.G. Cath		Michigan	Graduate Student
1955	P.M. Heertjes	Delft	Purdue	Lecturer
1955	H.R. van Nauta Lemke	Delft*	MIT	Research Scholar

Year	Scientist	NL University	US University	Position
1955	R.H. Nussbaum		Indiana	Research Scholar
1955	J.E. Prins	Delft	Berkeley	Research Scholar
1955	H. Reerink		Purdue	Graduate Student

* Van Nauta Lemke is listed in the Fulbright records as member of Delft Technical University at the time of his Fulbright grant, although he was actually an employee of Van der Heem.
Source: Fulbright Alumni 1949–1961, Dutch Fulbright Office (1990).
Note: some data about home institutions were unavailable, in which case the cell has been left blank.

PART 3

'Cold War' Science?

CHAPTER 8

The Absence of the East: International Influences on Science Policy in Western Europe during the Cold War

David Baneke

One of the problems of analyzing science in the context of something as large and complex as the Cold War is the difficulty of differentiating Cold War influences from other political or cultural factors. What distinguishes Cold War science from science that happened to happen during the Cold War? The politicization of trust and the efforts to gather intelligence or manage flows of information is one specific Cold War influence that has been convincingly identified, not least in several of the contributions to this volume. This article will focus on science policy and funding.

Throughout the Cold War, national governments were the dominant sponsors of scientific research. Science funding was, therefore, fundamentally a political issue. Decisions about the allocation of funds to specific areas were mostly made by scientists and government officials, but the overall level of funding, the structure of the funding institutions, the approval of specific large projects and decisions to join international efforts such as CERN, were decided at the highest political level. At this level, international politics was not just part of the context in which science policy developed: it was the central force. Any Cold War influence on science policy should be discernible at this level.

In recent scholarship, the story of Cold War science has become ever more complex and varied. It now investigates many disciplines beside the physical sciences, a variety of actors, and several different types of Cold War influences.[1] This diversity has enriched our understanding of the period, but it has also created an impression of the Cold War as an omnipresent influence on all aspects of scholarship, which reduces the analytical value of the 'Cold War' as a historical concept. In this article, I will look more closely at the role of Cold War politics in European science policy, with special focus on two aspects: firstly, in Western Europe, Cold War issues manifested themselves primarily in the relationship with the United States. This relationship was

1 Hunter Heyck and David Kaiser, "Focus: New Perspectives on Science and the Cold War, Introduction," *Isis* 101 (2010), 362–6.

 | DOI 10.1163/9789004264229_009

actively shaped by American and European politics at various levels, and only indirectly in reaction to Soviet actions.[2] Western European countries took the West, not the East, as point of orientation. Secondly, the Cold War was not the only international context that shaped European science and science policy in the post-war era. The global clash between an American-led Western bloc and an Eastern bloc dominated by the Soviet Union deeply influenced many aspects of science, but at the same time, European countries had to deal with, for example, the complex web of international relations within Europe, and with problems caused by decolonization. These issues also influenced each other: for example, the United States stimulated European cooperation. I will analyze the entanglement of global, regional and national politics, and assess their impact on Western European science policy, in order to present an analysis of Cold War era science policy which considers the Cold War as one part of a more complex political context.

Europe's American Focus and the Absence of the East

It is remarkable that in much of the literature about Cold War science in Western Europe, including many of the papers in this volume, the Soviet Union and the Eastern Bloc are nearly invisible. They are present in the background, but their actions do not seem to influence the main narrative in any direct way.[3] It may seem strange that one can write about a war, even a cold one, without looking at one of its main participants, but on second thought, there are good reasons for the absence of the East from Cold War narratives. From the perspective of many of the smaller countries on either side of the Iron Curtain, only the actions of 'their' superpower mattered in each

2 Much literature focuses on American policy, and considers Europe a relatively passive actor. Several authors of a recent *Centaurus* issue on Cold War Science commented on this bias, including John Krige, "Building the Arsenal of Knowledge," *Centaurus* 52 (2010), 280–96, although Krige still focuses mostly on the American side in that article. Friso Hoeneveld and Jeroen van Dongen, "Out of a Clear Blue Sky? FOM, the Bomb and the Boost in Dutch Physics Funding after World War II," *Centaurus* 55 (2013), 264–93, present an interesting case from a European perspective.

3 David Caute, "Foreword," in Hans Krabbendam and Giles Scott-Smith (eds.), *The Cultural Cold War in Western Europe 1945–1960* (London, 2003), pp. vii–ix, commented on the 'vanishing' of the Soviet Union from Cold War cultural studies, attributing this phenomenon to revisionist, New Left interpretations of the Cold War, as instigated by the aggressive, almost imperialist policy of the United States.

country's daily politics and practice. For Western Europe, that superpower was the United States.

The resulting picture is well-known in political history. By 1950, when NATO had been founded, the most immediate communist threats in France and Italy had been avoided and a stable status quo had been established in Germany, defense against a Soviet invasion was no longer the most urgent problem for European politicians, even though the fear of the Soviets was still real. Once the nuclear arms race escalated in the 1950s, most European countries had to conclude that they could not significantly contribute to any actual military confrontation between the superpowers: their most likely role would simply be that of battlefield. Even Britain and France could not hope to compare their capabilities to the arsenals of the United States or the Soviet Union in any serious manner.

The powerlessness of Europe may have seemed desperate, but that was not how it was perceived by the Europeans, especially after the 1960s. Psychologically, fighting the Cold War was increasingly viewed as an American affair. One might say that Europe had outsourced its protection against the East to the United States. Several historians have pointed out that the very powerlessness of Europe was advantageous in a way. Tony Judt finds that the growing emphasis on nuclear weapons and intercontinental missiles released European states from the requirement to compete in an arena in which they could not match the superpowers' resources. "So they cultivated their gardens instead."[4] For most Western European countries, the post-war decades were a period of social stability, fast economic growth and increasing prosperity. It was the age of the German *Wirtschaftswunder*, the French *trente glorieuses* (despite the turbulent politics of the Fourth Republic) and the Italian *miracolo economico*. Even Franco's Spain experienced *el milagro español* in the 1960s.

The 'cultivation of European gardens' was not necessarily as quiet and pastoral as this metaphor may suggest. The post-war period was also a period of complex international politics on many levels. Three important issues were summarized by Lord Ismay, the first Secretary General of NATO, who famously defined the goals of NATO as "keeping the Russians out, the Americans in and the Germans down." Initially, the third of those goals was the most urgent for many European countries. The repercussions of World War II continued to shape European politics for many years. Even after a stable equilibrium had been established on the European continent, intra-European relations remained complex and dynamic. European countries were constantly trying to find new ways to compete and to cooperate. Soon an intricate network of

4 Tony Judt, *Postwar* (London, 2005), pp. 247, 256.

international associations and organizations came into existence, at times with overlapping mandates. Many of these organizations were directly or indirectly sponsored by the United States.

In the meantime, the second goal, of keeping America involved in Europe, was also a pressing one. One reason for the urgency was that the Americans' involvement would, in turn, take care of the first goal, of keeping out the Russians, by providing a serious deterrent against the Soviet Union. Europe could not provide that deterrent itself, while fear of the Communist threat was real. For that reason, it was important to maintain a sizeable military force: in order to combat the Russians directly, but even more so in order to show America that Europe was doing its part. It was crucial to keep the United States interested in protecting Europe. But there were other, additional reasons for cultivating connections with the United States. The most significant among them was the desire to be granted access to American scientific, technological and advanced project management knowledge, as well as American currency and funds. Finally, the smaller European countries needed America to remind France and Germany that they were not the most intimidating presences in European politics—a slightly modified version of the third goal of NATO. All reasons remained valid throughout the Cold War, even though many aspects of transatlantic relations changed in the 1970s and 1980s.

America's role as protector of Western Europe and its security provided it with both an enormous influence in, and an interest in influencing, Europe. America desired to keep Western Europe firmly on its side in the global struggle. Thus, the USA and Western Europe were mutually dependent, albeit in a rather asymmetrical way. As John Krige has shown in *American Hegemony*, this relationship had direct consequences for science.[5] In the first few post-war years, American policy shaped European science, creating an institutional infrastructure based on American models. After the establishment of this infrastructure, American influence remained strong.

Lord Ismay's pithy summary of NATO's goals may be supplemented with another fundamental international issue of note, in which NATO did not have a direct role: several European countries struggled with the deconstruction of their colonial empires. Decolonization was a long and painful process, first and foremost for the newly-independent people, but also for the former colonial powers. This struggle influenced the relations between European countries and the United States, since the US supported independence in several cases (not least because of Cold War rationales, as the US wanted the new countries

---

5 John Krige, *American Hegemony and the Postwar Reconstruction of Science in Europe* (Cambridge, MA, 2006).

to become its allies). Moreover, European countries were able to spend substantial resources on military causes in their colonies because it was assumed that European security was guaranteed by the US. When the Netherlands sent almost its entire army to Indonesia in 1946–1949, the country was left nearly defenseless. The United States objected, insisting that European countries should contribute more to their own protection.[6]

Overall, European politicians and diplomats had to simultaneously act on several international playing fields, addressing the Communist threat, maintaining good relations with America, dealing with decolonization, and navigating a complex intra-European political web in which both cooperation and competition were essential. These elements all influenced each other in complex ways. To understand European science policy, therefore, this political big picture is essential.[7]

Scientific Manpower: The Technology Gap

Paul Forman and David Kaiser, among others, have analysed the enormous increase of science spending in the United States in the 1950s and 1960s. They convincingly explain this surge in terms of Cold War developments: the expenses were intended to maintain world leadership in terms of scientific knowledge, technological capability, and manpower in science-related fields. The political discourse surrounding research funding was dominated by real or perceived 'gaps' in relation to the Soviet Union, including a shortage of manpower.[8]

Science departments at Dutch universities, too, enjoyed years of abundance in the 1950s and 1960s. Money was plenty, talented students abounded,

---

6 Wim Klinkert and Gerke Teitler, "Nederland van neutraliteit naar bondgenootschap," in Bob de Graaf, Duco A. Hellema and Bert van der Zwan (eds.), *De Nederlandse buitenlandse politiek in de twintigste eeuw* (Amsterdam, 2003), pp. 9–36, especially p. 24; Duco A. Hellema, *Dutch Foreign Policy: The Role of the Netherlands in World Politics* (Dordrecht, 2009), especially pp. 142–3.

7 Cf. Gérard Bossuat, "Les coopérations européennes pour la recherche scientifique et technique," *Journal of European Integration History* 12 (2006), 5–10.

8 Paul Forman, "Behind Quantum Electronics: National Security as Basis for Physical Research in the United States, 1940–1960," *Historical Studies in the Physical and Biological Sciences* 18 (1987), 149–229; David Kaiser, "Cold War Requisitions, Scientific Manpower, and the Production of American Physicists after World War II," *Historical Studies in the Physical and Biological Sciences* 33 (2002), 131–59; David Kaiser, "The Physics of Spin: Sputnik Politics and American Physicists in the 1950s," *Social Research* 73 (2006), 1225–52.

and staff numbers were increased every year. Scientists from that period still remember with some amazement the time when nearly every funding application was granted. Historical accounts of the funding boom often refer to a report of the 'Casimir committee.' Chaired by the influential physicist and research director of Philips Electronics, Hendrik Casimir, this committee recommended a significant increase in research funding. Casimir presented his report in October 1958, exactly one year after Sputnik, as is often noted. Even those who note that the creation of the committee was not directly motivated by Sputnik—because the preparations had started earlier—still argue that Sputnik scared politicians into implementing all the recommendations without any hesitation.[9]

Remarkably, however, none of the relevant documents referred to Sputnik or a 'manpower gap', and neither did any of the reports on discussions in politics and university boards concerning the need for increased science funding, which took place shortly after the publication of the report.[10] The Soviet Union or Eastern European countries were not mentioned, except for one reference in passing to the fact that there was some concern in NATO about the number of excellent students in Eastern Europe; the committee reported this without endorsing it as an argument for their recommendations.[11] In all of these documents, economic considerations and comparisons with other Western countries dominated the argumentation (it is obviously impossible to know what was said off the written record). But it is significant that 'Sputnik' (or 'Spoetnik' in the Dutch spelling) was only mentioned 19 times in the proceedings of Dutch parliament (*Tweede Kamer*) before 1980. Most of those references were made by members of the Communist party; several others by politicians who downplayed the significance of the satellite.[12] As one member of parliament belonging to the Catholic People's Party said during the debate on the 1958

9 For example: Albert Kersten, *Een organisatie van en voor onderzoekers: ZWO 1947–1988* (Assen, 1996), p. 158; Ernst Homburg, *Speuren op de tast: een historische kijk op industriële en universitaire research* (Maastricht, 2003), p. 43; Ton van Helvoort, *De KNAW tussen wetenschap en politiek: de positie van de scheikunde in de Akademie in naoorlogs Nederland* (Amsterdam, 2005), pp. 81–2, 92; Klaas van Berkel, *De stem van de wetenschap: geschiedenis van de Koninklijke Nederlandse Akademie van Wetenschappen*, vol. 2 (Amsterdam, 2011), pp. 311–3.

10 David Baneke, "De vette jaren: de Commissie-Casimir en het Nederlandse wetenschapsbeleid 1957–1970," *Studium* 5 (2012), 110–27.

11 [Hendrik B.G. Casimir], *Voorzieningen ten behoeve van de research binnen de faculteiten der wis- en natuurkunde der Nederlandse universiteiten* (The Hague, 1958), p. 10.

12 The Proceedings can be searched online at www.statengeneraaldigitaal.nl [accessed 16 June 2014].

defense budget, "the strategic balance between East and West had been maintained without major change, even after the launch of a [sic] sputnik."[13] Clearly, there was no 'Sputnik shock,' at least not on the political level.

The Casimir report had been triggered in December 1956 by a joint complaint, by the heads of the science departments of the universities of Leiden, Groningen and Utrecht, about the consequences of a government-wide hiring freeze. The protest was initiated by the astronomer Jan Oort, then dean of the science department in Leiden, and was well-timed, as it coincided with debates within the education ministry about the future of the Dutch higher education system. Education minister Jo Cals was working on a fundamental overhaul of the entire education system, and the national bureau of statistics had just issued a report predicting an exponential increase of student numbers at the universities up to 1970; their numbers had been growing steadily since World War II. This increase was not caused by a GI bill or a campaign to educate and recruit future scientists for military research, but mainly by a socio-economic development: an increasing number of middle class youths attended university. Furthermore, the population of the Netherlands was growing fast, and especially the children of the post-war baby boom were expected to enter university in large numbers by 1970. This caused a favorable political climate for the increase of university budgets.[14]

Might the Cold War have been the reason why so many of the new students, and so much of the funding, went to the science departments instead of, say, law or economic science departments?[15] There were obvious reasons why science and technology were favored, but direct competition with the Eastern Bloc was rarely mentioned. The Cold War may still have played a role, then, but apparently in an indirect way, within broader economic and diplomatic concerns. In the 1950s, some of the primary concerns of Dutch policy makers—scientists, politicians, university administrators and representatives of industry—were the economic reconstruction of the Netherlands and a perceived technological gap, not with the Eastern Bloc, but with other Western countries, especially America and Britain. After World War II there was a strong sense that those countries had leapt ahead in science and technology,

---

13 Eddy Visch (KVP): "het strategisch evenwicht tussen Oost en West [is] ongeschokt gehandhaafd gebleven, ook na het oplaten van een spoetnik," Handelingen van de Tweede Kamer, 11 February 1958, available at www.statengeneraaldigitaal.nl [accessed 16 June 2014].

14 Jan Willem Brouwer and Peter van der Heiden (eds.), *Het kabinet-Drees IV en het kabinet-Beel II, 1956–1959: het einde van de rooms-rode coalitie* (The Hague, 2004); Peter Baggen, *Vorming door wetenschap: universitair onderwijs in Nederland 1815–1960* (Delft, 1998).

15 Patricia Faasse, *De Utrechtse bètawetenschappen 1815–2011* (Hilversum, 2012).

while innovation and research in the Netherlands had stagnated during the German occupation.

The desire to keep up with the leading Western nations was motivated by economic as well as political and military considerations. After the War, Dutch economists concluded that heavy industry and agriculture provided few opportunities for such a small country with such few natural resources as the Netherlands. Post-war reconstruction of the Dutch economy therefore needed to focus on industry based on knowledge which added value in expertise rather than resources, especially high tech industry as represented by Philips, airplane manufacturer Fokker, and the chemical industry. For this purpose the country needed to train more scientists and engineers. The fear of falling behind internationally persisted for several decades.[16] The sentiment also had a political aspect, which was expressed, for example, by Gerard Kuiper, a Dutch astronomer who had emigrated to the US in the 1930s. When he returned to Europe in 1945 as a member of the Alsos mission, he advised Willem Schermerhorn, who had recently been appointed Dutch prime minister and was an engineer himself, of the need for immediate action. America and Britain had monopolized large parts of science and technology, Kuiper wrote. If the Netherlands were not to establish connections with them soon, "I'm afraid the US will think of Holland as of Portugal or Romania [...]. The US already tends to think of Europe as a 'quantité négligeable'."[17] Investing in science would positively affect national morale and the Dutch economy, and was necessary for the general relationship with the United States, and thus ultimately for the Dutch position in international politics and the country's national security.

It was no coincidence that the committee on research funding was chaired by H.B.G. Casimir, the director of Philips' impressive research establishments. The company was then the prime example of national economic achievement, and it was an arsenal of strategic knowledge. Notably, and at the same time, Casimir was also involved in NATO's attempts to create a science policy. Another member of the Casimir committee was H.W. Slotboom, the director of the Shell Laboratories in Amsterdam. Together, Casimir and Slotboom

16 Kees Schuyt and Ed Taverne (eds.), *1950. Welvaart in zwart-wit* (The Hague, 2000), pp. 120–5; Peter Baggen, Jasper Faber and Ernst Homburg, "The Rise of a Knowledge Society," in Johan Schot, Harry Lintsen and Arie Rip (eds.), *Technology and the Making of the Netherlands: The Age of Contested Modernization, 1890–1970* (Zutphen, 2009), pp. 253–323.

17 Gerard Kuiper to Jan Hendrik Oort, 15 August 1945, Leiden University Library, J.H. Oort Papers, inventory number 155b.

represented the largest and most powerful companies, and their institutions were by far the largest employers of physicists and chemists in the country. The other four members of the committee were university professors: J.H. Oort (astronomy, Leiden), C.J. Gorter (physics, Leiden), J.Th.G. Overbeek (physical chemistry, Utrecht), and W.H. Arisz (plant physiology, Groningen). All committee members had established strong American connections. The committee solicited advice from prominent Dutch scientists abroad, including Gerard Kuiper (astronomy, Tucson), Nicolaas Bloembergen (physics, Harvard) and Nikolaas Tinbergen (ethology, Oxford). Their advice on research management clearly influenced the final report, for example, it lead to the inclusion of the recommendation that academic institutions should have professional managers, instead of being managed by the professors.

The Casimir committee's report mainly addressed the questions of how the economic potential represented by talented students might be used in an optimal manner, and of how one might close the technology gap between the Netherlands and the leading Western countries. Almost all of the report's recommendations were implemented. In the 1960s, science departments became the largest departments of their respective universities in terms of student numbers and funding. After 1970 the growth rates leveled off, and other departments, especially Social Sciences, grew relatively larger.[18]

Direct competition with America was naturally out of the question, but the aim was to 'catch up and keep up' as much as possible. There was some uncertainty about the merit and feasibility of competition with large European countries such as Britain, France and Germany, but it was tried nonetheless. In America, political arguments for increased science funding were mostly based on comparisons with the Soviet Union. In the Netherlands, it was more effective to point at the danger of being left behind by the United States.

Limits and Gifts

The soft power of American technological primacy was combined with its more overt political power. In practice, America was able to define boundaries to Europe's freedom of action. For example, in their monumental history of CERN, Krige and Pestre describe how CERN could only be founded after the American government gave its permission. Permission was granted after the Soviet nuclear bomb was tested in 1949: it was obvious then that America's

18 Faasse, *De Utrechtse bètawetenschappen.*

nuclear monopoly was untenable.[19] This sequence of events was characteristic of the era: Soviet actions provoked an American reaction, and the subsequent American action, in turn, influenced European actions. This two-stage structure illustrates the nations' different perspectives on international politics. America monitored the choices of the Soviet Union, while Europe watched the United States.

CERN was not an American project, but America was a constant presence, limiting the spectrum of possible actions, including the choice of research topics and the flow of information. From the European perspective, one of the main goals of CERN was to elevate European science to a higher level, in order to be able to compete with America. Competitors were not imagined in the East, but rather across the Atlantic, such as Brookhaven and, later, Fermilab.[20] The same holds true for the European Southern Observatory, which was also founded in the 1950s. The structure of this organization was explicitly modeled after CERN, and the two organizations' officials coincided in part. The foundation of ESO was expedited by a large grant from the American Ford Foundation, but its explicit goal was to compete with the large American telescopes in California.[21]

Eric Hobsbawm points out that the European Community was, "like so many other things in post-1945 Europe, created both by and against the USA."[22] The same was the case for CERN and ESO, as well as that for European space organizations. As Robert Smith writes, space science was "a gift of the Cold War."[23] It started with upper atmosphere research with sounding rockets, an activity which, in most countries, was conducted by the military.[24] Much space technology was developed for military and intelligence purposes, and all launch

19 Armin Hermann, John Krige, Ulrike Mersits and Dominique Pestre, *History of* CERN, vol. 1: *Launching the European Organization for Nuclear Research* (Amsterdam, 1987), esp. pp. 71–2, 89.

20 Hermann et al., *History of* CERN, vol. 1, pp. 529, 537; Armin Hermann, John Krige, Ulrike Mersits, Dominique Pestre and Laura Weiss, *History of* CERN, vol. 2: *Building and Running the Laboratory* (Amsterdam, 1990), pp. 719, 789–91.

21 Adriaan Blaauw, *ESO's Early History: the European Southern Observatory from Concept to Reality* (Garching, 1991); Claus Madsen, *The Jewel on the Mountaintop: The European Southern Observatory through Fifty Years* (Garching, 2012).

22 Eric Hobsbawm, *The Age of Extremes 1914–1991* (London, 1994), p. 240.

23 Robert W. Smith, "The Making of Space Astronomy: A Gift of the Cold War," in Alison D. Morrison-Low, Sven Dupré, Stephen Johnston and Giorgio Strano (eds.), *Earth-Bound to Satellite: Telescopes, Skills and Networks* (Leiden, 2011), pp. 235–49.

24 David DeVorkin, *Science with a Vengeance: How the Military Created the US Space Sciences after World War I* (New York, 1992).

vehicles were based on ballistic missile designs. But the most significant gift of the Cold War was the American creation of NASA as a civilian space agency with a sizeable science mandate. This was a direct reaction to Sputnik, shaped by Cold War considerations.[25]

While in Europe, space science also started after Sputnik, European actions were not shaped by the Soviet actions, but by the American reaction to them, as had been the case for CERN. In his book about the German space program, Niklas Reinke accuses the Adenauer administration of failing to grasp the implications of the flight of Sputnik and of acting too slowly: Adenauer's mistake, according to Reinke, was the reason why German space policy was fairly passive, following in the footsteps of international developments rather than being directed forward by well-defined German policy aims.[26] However, the same developments could be observed in several other Western European countries. European military men, politicians, intellectuals, and certainly also scientists were impressed by the Soviet satellite.[27] But in most European countries, setting up a national space program was not a realistic option. Spaceflight was the domain of the superpowers. Only France and Britain were working on launchers, with varying success, in an attempt to claim superpower status.

Outer space only came within the reach of smaller countries as a result of international political developments, more specifically because of American actions that, in turn, were guided by the American desire to regain momentum after the humiliating blow of Sputnik, and the even more embarrassing failure to launch the Vanguard rocket a few months later. The United States government took the initiative to establish two international forums to discuss space politics: the United Nations Committee on Peaceful Use of Outer Space, and the Committee on Space Research (COSPAR). These organizations invited countries and scientists to report on their plans and actions. The UN committee mostly focused on legal issues related to space flight, while COSPAR was a scientific organization.

At the second meeting of COSPAR, in March 1959 in The Hague, the American representative Richard Porter told the surprised assembly that his government offered to launch foreign scientific satellites on American rockets for free. The

25 Walter MacDougall, *The Heavens and the Earth: A Political History of the Space Age* (New York, 1986); Roger D. Launius, John M. Logsdon and Robert W. Smith, *Reconsidering Sputnik: Forty Years Since the Soviet Satellite* (Newark, NJ, 2000).

26 Niklas Reinke, *The History of German Space Policy: Ideas, Influences, and Interdependence 1923–2002* (Paris, 2007), pp. 134–5.

27 Igor J. Polianski, Matthias Schwartz (eds.), *Die Spur des Sputnik: Kulturhistorische Expeditionen ins Kosmische Zeitalter* (Frankfurt a.M., 2009).

offer was made possible by the newly founded NASA. This offer was a classic case of using science as a diplomatic tool. America aimed to present itself as a benign and benevolent hegemon that made its technology available for the benefit of mankind, in contrast to the secretive and militaristic Soviet Union. Less openly, it also wished to exert as much influence as possible on foreign space programs, preferably without appearing to interfere.[28] The American offer enabled European countries to develop a space science program without having to wait for the development of a European launcher. For national, intra-European as well as transatlantic political reasons, European countries accepted the offer and used it to create both national space programs as well as a European space research organization.

Initial ideas for European cooperation in space were discussed during COSPAR meetings. These discussions involved many of the founders of CERN. Eventually, this led to the foundation of the European Space Research Organization (ESRO), modeled after CERN. Just as in the case of CERN, it was American Cold War politics that made European action possible.

Around the same time, Britain decided to convert its military missile ('Blue Streak') into a civilian launcher. More specifically, it was suggested that the missile might be used by a future European space organization as the basis for a new launch vehicle.[29] The decision was made by British politicians, but it would not have been possible without approval of the US, which had supported the British missile program, and which also developed a missile that rendered the British one redundant for military use. It was certainly in America's interest that foreign space programs be civilian, not military, in nature.[30] Lorenza Sebesta has argued that the British offer to Europeanize its missile program was shaped by a complex interplay between the United States, Britain and France, each of which had different ideas about European cooperation in launcher technology. This was related, among other things, to the French ambition to

28 John Logsdon, "The Development of International Cooperation," in John Logsdon (ed.), *Exploring the unknown*, vol. 2 (Washington D.C., 1996), pp. 1–16; John Krige, "Technology, Foreign Policy, and International Cooperation in Space," in Steven J. Dick and Roger D. Launius (eds.), *Critical Issues in the History of Spaceflight* (Washington D.C., 2006), pp. 239–60.

29 Lorenza Sebesta, "Choosing Its Own Way: European Cooperation in Space, Europe as a Third Way between Science's Universalism and US hegemony?" *Journal of European Integration History* 12 (2006), 27–55; Sir Harry S.W. Massey and Malcolm O. Robins, *History of British Space Science* (Cambridge, 1986); John Krige, Arturo Russo and Lorenza Sebesta, *A History of the European Space Agency 1958–1987* (Noordwijk, 2000).

30 Krige, "Technology, Foreign Policy;" cf. Arnold Frutkin, *International Cooperation in Space* (Englewood Cliffs, NJ, 1965).

develop an independent European launcher; Britain's ambition to firm up relations with the European continent while retaining its 'special relationship' with America, and the US policy concerning knowledge transfer to France via Britain.[31] The final outcome of these negotiations was the foundation of ELDO, the European Launch Development Organization, with significant British and French contributions, and with America's blessing.

The American offer to launch foreign satellites and the British offer to Europeanize its missile project prepared the way for the two European space organizations, ESRO and ELDO. The organization of launcher development and space science in separate institutions was, of course, a direct consequence of Cold War constraints. Rockets were politically too sensitive to mix with supposedly neutral science—in Europe, at least. It was only in 1975, after a series of political and financial crises which especially affected ELDO, that the two organizations were merged into the new European Space Agency (ESA).[32]

National Space Programs in Europe

In Germany, the Netherlands, and most other European countries that developed a space program, discussions about European organizations initiated the start of national space programs as well.[33] Each country had its own reasons to join or not join ESRO and ELDO. A brief survey of several countries' motivations will provide an illustration of the complex web of national and international interests that shaped national space policies. They also executed much influence on the development of disciplines such as astronomy, geophysical science and aerospace technology in their individual national manifestations.

For Germany, ESRO and ELDO provided opportunities for involvement in a sensitive but significant technological field, and for supporting German aerospace industry, without violating the restrictions that were imposed by the victors of World War II.[34] On a political level, Germany also sought to break out of

---

31 Sebesta, "Choosing its own way."

32 Krige and Russo, *A History of the ESA*.

33 Reinke, *The History of German Space Policy*; David Baneke, "Space for Ambitions: The Dutch Space Program in Changing European and Transatlantic Context," *Minerva* 52 (2014), 119–40; Stephan Zellmayer, *A Place in Space: The History of Swiss Participation in European Space Programmes, 1960–1987* (Paris, 2008), pp. 23–30, provides an overview of the literature about European space programs.

34 Anke Marei Ludwig, "Platz gefunden.—Ziele klar? Die Politik der europäischen Mitgliedstaaten im NATO-Wissenschaftsausschuss (1957–1967)," *Journal of European Integration History* 12 (2006), 91–105.

its post-war isolation: the country tried to establish itself as a 'broker' between French and American interests in the European continent, and between large and small countries within Europe.[35] The Swiss showed somewhat similar interests, except that their restrictions were self-imposed. Supporting ESRO, like joining CERN, offered Switzerland a way to demonstrate its willingness to cooperate internationally and to contribute to international projects without violating its strict policy of neutrality.[36] For both Germany and Switzerland, the apolitical and non-military nature of the involved organizations was crucial, just as it was important for NASA to be perceived as a civilian organization.

In Norway, however, nearly all space activities were military in nature. Norwegian space research focused on investigating the upper atmosphere, with the goal of improving communications in the polar region. John Collett describes Norwegian policy in this matter as fairly passive: America took the lead in Norwegian space research, and so much so, that Norway even declined to join the European space organizations. The US had already provided it with all necessary materials. Norway later applied for ESA membership in the 1970s, when it became clear that a much-needed American sea surveillance system would not be made available.[37] Altogether, American policy shaped Norwegian actions directly.

France was, once more, a special case. France was perhaps the only European country with an active space policy in the early years of space flight. It aimed to claim great power status in its own right, independent from America, and it attempted to enlist other European countries in space projects in order to gain their support for European independence in spaceflight. France did not wish to be reliant on America for the launch of satellites.[38] Commercial interests were also significant, since France boasted a substantial aerospace industry.

These examples illustrate that in many European countries, local and regional considerations shaped science policy as much as global politics, even with regard to international scientific cooperation. The developments in Italy, Belgium, and Sweden lend themselves to similar analyses. Anke Marei Ludwig

35 Reinke, *The History of German Space Policy*.

36 Zellmayer, *A Place in Space*; Bruno J. Strasser, "The Coproduction of Neutral Science and Neutral State in Cold War Europe: Switzerland and International Scientific Cooperation," 1951–69', *Osiris* 24 (2009), 165–187.

37 John P. Collett (ed.), *Making Sense of Space: The History of Norwegian Space Activities* (Oslo, 1995), p. 285.

38 Lorenza Sebesta, *The Availability of American Launchers and Europe's Decision To Go It Alone* (Noordwijk, 1996).

has argued that even the discussions in NATO about large scientific projects were shaped by inter-European politics.[39]

In the following section I will briefly discuss one last example, the Netherlands, which is the case with which I am most familiar. For the Netherlands, one major reason to join ESRO and ELDO was that it followed the lead of France, Germany and Britain in order to keep up with them, especially since this was such a highly visible and prestigious field of research. The Netherlands never fully accepted its role as a small country, instead preferring to act as a rather small large country.[40] Similar considerations motivated Dutch efforts in fields such as nuclear science and advanced communications technology. In each case, the effort needed to be substantial enough to be taken seriously on the international stage, and to win a seat in critical international meetings.[41] Similarly, the Dutch defense budget was in part intended to secure its position within NATO.

The fact that Britain was involved in the European space organizations made the situation especially interesting. The Netherlands was always urging Britain to become more involved in European affairs, as a counterweight to France. In this sense, Britain had a similar role to that of America: the role of an outsider who could keep the larger European countries in check. Another important motivation for a Dutch space program was the presence of powerful electronics (Philips) and aircraft (Fokker) industries that hoped to enter this new and potentially lucrative field.

The American offer to launch satellites not only kick-started European space programs, both nationally and internationally, but it also played an important role in inter-European competition. ESRO happily accepted the American offer to launch its satellites, but individual countries did the same, so that they could launch satellites that were built outside of the European framework. These national programs were partly used to compete for contracts and positions of influence within ESRO.

---

39 Ludwig, "Platz Gefunden;" cf. Sebesta, "Choosing Its Own Way."

40 Alfred E. Pijpers, "Dekolonisatie, compensatiedrang en normalisering van de Nederlandse buitenlandse politiek," in Nicolaas C.F. van Sas (ed.), *De kracht van Nederland: Internationale positie en buitenlands beleid in historisch perspectief* (Haarlem, 1991), pp. 204–18; Joris J.C. Voorhoeve, *Peace, Profits and Principles: A Study of Dutch Foreign Policy* (Leiden, 1985), pp. 10–1.

41 Cf. Jaap van Splunter, *Kernsplijting en diplomatie: de Nederlandse politiek ten aanzien van de vreedzame toepassing van kernenergie, 1939–1957* (Amsterdam, 1993); Hoeneveld and Van Dongen, "Out of a clear blue sky?"

The Netherlands took up the American offer, in the 1960s to launch the Astronomical Netherlands Satellite, and in the 1970s to launch the Infrared Astronomical Satellite. These bilateral Dutch-American projects were intended to strengthen the position of the Netherlands in the European context. Economic competitiveness of the Netherlands was the main motive of these projects. Philips and Fokker, two national flagship companies, wished to demonstrate their capabilities, in order to be able to compete seriously for ESRO and ELDO contracts. Interestingly, acquiring advanced project management knowledge was also an important motive. The companies wished to acquire the necessary skills for managing extensive development projects, a field in which NASA excelled. The acquisition of these skills would be achieved in direct cooperation with NASA. The Ministry of Economic Affairs supported these ambitions, and provided funding for two successive Dutch-built satellites.[42]

These considerations were of little relevance to astronomy, but for various reasons an astronomical satellite was considered the best vehicle for realizing the mentioned ambitions. Science was a convenient vehicle for economic policy: it was considered politically neutral and carried a large public appeal. More specifically, Dutch astronomy was a well-organized community with an excellent international reputation.[43] The Dutch government and industry approached astronomers directly to ask if they could find use for a satellite—as indeed they did. As a result, the Astronomical Netherlands Satellite was built (ANS, launched in 1974), followed by the American-Dutch Infrared Astronomical Satellite (IRAS, launched 1983).

In this case, American Cold War policy facilitated a course of action that would otherwise have been impossible. Indirectly, the scientific satellites were 'gifts of the Cold War,' because the NASA offer which had made these projects conceivable was motivated by Cold War considerations. At the same time, Cold War politics defined clear boundaries between achievable goals and plans that were impossible to translate into practice. But regional politics and economic strategy were the most direct motives for the Dutch government to support satellite projects.

42 Baneke, "Space for Ambitions;" Niek de Kort, *Ruimteonderzoek: de horizon voorbij* (Amsterdam, 2008).

43 David Baneke, *De ontdekkers van de hemel. De Nederlandse sterrenkunde in de twintigste eeuw* (Amsterdam, 2015).

Europe as a Third Power?

If the smaller European countries did not play a significant role in the great global power struggle in the 1950s and 60s, their situation could be turned into their advantage. The trope of small countries acting as mediators between great powers—a role that seemed natural since they posed no military threat on their own—was well-established in the early twentieth century. Belgium, the Netherlands, Switzerland, and the Scandinavian countries all tried to establish an international significance in this way. Science played an important role in the image they projected, as witnessed by the creation of Nobel Prizes and the Solvay Conferences.[44]

This dynamic did not disappear after World War II. As we have seen, Switzerland used international scientific organizations to underline its politics of 'neutrality and solidarity.' But even in NATO-member countries a sense of neutrality could remain. When asked about the relative prominence of Dutch officials in international astronomical organizations such as COSPAR, ESO and the International Astronomical Union, several leading Dutch astronomers mentioned Dutch 'neutrality' during the Cold War, which marked them as suitable candidates for positions which the superpowers did not want to concede to each other.[45] They clearly knew that the Netherlands was not a neutral power in any political sense, but statements of neutrality illustrate that they considered the Cold War a matter beyond their immediate concern—even though, as astronomers and space scientists, they benefitted from the Cold War more than anyone else.

The relationship between European countries and the United States changed in the 1970s and 1980s. While America was increasingly preoccupied with domestic problems and the Cold War entered a phase of détente, Europeans became increasingly critical of American policy, and transatlantic relations grew to be more complicated. European students and intellectuals started questioning American hegemony, even while they adopted American popular culture. At the same time, NATO's change from 'massive retaliation' to 'flexibility in response' as its guiding strategy required a more active role for

44 Rebecka Lettevall, Geert Somsen and Sven Widmalm (eds.), *Neutrality in Twentieth-Century Europe. Intersections of Science, Culture, and Politics after the First World War* (London and New York, 2012); Willem Otterspeer and J. Schuller tot Peursum-Meijer, *Wetenschap en wereldvrede. De Koninklijke Akademie van Wetenschappen en het herstel van de internationale wetenschap in het Interbellum* (Amsterdam, 1997).

45 Interviews by the author with astronomers Cornelis de Jager (May 2010), Harry van der Laan (January 2012), and Hugo van Woerden (February 2012).

Europe in military matters, while the West-German *Ostpolitik* led to a détente in German-German relations.[46]

In the 1980s, European confidence in national science and technology increased, even though American primacy was still uncontested in most fields. For the first time since World War II, Europe was able to claim a serious part in international competition in some highly visible scientific fields and technologies. The most dramatic example was the Ariane launcher, which provided Europe with a credible launch capability for the first time. Especially after the Space Shuttle Challenger disaster of 1986, Ariane became a true alternative to American launchers. Other examples of successful European science include the discovery of the W and Z bosons and the inauguration of the Large Electron–Positron Collider at CERN. Around the same time, European organizations decided to build the largest telescope in the world (ESO's Very Large Telescope, VLT), and an even more powerful accelerator (CERN's Large Hadron Collider, LHC). With Spacelab (1983–1988), Europe even developed a manned space program, the pinnacle of showcase technology.

Remarkably, even when Europe started to challenge American hegemony in this way, America kept sponsoring European efforts in science. Spacelab, for example, was still completely dependent on an American infrastructure (the Space Shuttle). Naturally, American primacy was still recognized in Europe. In their recent book on NASA's international cooperation, Krige, A.L. Callahan and A. Maharaj refer to a British official's evaluation of the European participation in the American (later: International) space station, according to whom the question was not whether it made sense to build a space station, but, "[g]iven that the US has decided to build one, and has invited us to join, can we afford not to?"[47]

Conclusion

Many considerations shaped European science policies during the Cold War. The great power struggle with the East was one of them, but not always the most urgent one. To understand the impact of the Cold War on science, we have to assess its importance carefully and in comparison with other factors. It would be incorrect to ascribe the post-war surge in science funding solely to the Cold War. Even in space research, one of the most iconic Cold War activities, national and regional considerations played a significant role.

46 Klinkert and Teitler, "Nederland van neutraliteit," p. 28.

47 John Krige, Angelina Long Callahan and Ashok Maharaj, *NASA in the World: Fifty Years of International Collaboration in Space* (New York, 2013), p. 251.

Moreover, from a European perspective, the Soviet threat was not a direct concern in this context. In several ways, the Cold War was regarded as an American affair. The Cold War had a cascading effect: Cold War considerations were extremely important in American policy making, which, in turn, defined the limits of European policy options. American policy created opportunities and stimulated specific research in Europe, but it also blocked certain courses of action.[48]

The impact of the focus on the USA in European science policy was varied. America was able to regulate flows of information in several cases, including high-profile scientific projects such as CERN. Certain disciplines certainly benefitted from their symbolic role in the Cold War struggle, and from technologies that had originally been developed for military purposes. Space science was a prime example. America's Cold War diplomacy made it possible for European countries to embark on space projects two decades before a reliable European launcher became available. Apart from its direct influence, cooperation with the United States was also used by European countries in their competition with each other. Finally, the global political struggle, in which nearly every subject was polarized, also created the need for a politics-free, neutral zone, both for diplomatic purposes and for economic and technological competition in isolation from political sensitivities. Science provided such a zone, and this created attractive possibilities for scientists, especially for those from small countries.

Many of the dynamics I have described in this article remained in place after the end of the Cold War. Even when Europe was no longer dependent on America for protection against the Soviet Union, America remained a powerful actor in European politics. America was able to retain its primacy in many fields, including science and technology. It remained the main sponsor of international space activities, and, more generally, it provided the standard by which Europeans assessed their scientific and technological activities.[49]

The picture that I have painted here provides a more complex model of Cold War science and American hegemony than previously established. International relationships were dynamic on all levels: national, regional, and global. The different levels interfered with each other, and all participants were actively trying to shape them according to their own interests. The one actor that was absent from many of the considerations was the Soviet Union.

---

48 I would like to thank Christian Joas for the term 'cascading influence.'

49 Cf. Joris van Eijnatten, Toine Pieters and Jaap Verheul, "Big Data for Global History: The Transformative Promise of Digital Humanities," *BMGN-Low Countries Historical Review* 128 (2013), 55–77, on 'reference cultures'.

CHAPTER 9

Colonial Crossings: Social Science, Social Knowledge, and American Power from the Nineteenth Century to the Cold War

Jessica Wang

Studies of the Cold War have generally treated World War II as the pivot point of twentieth-century history, from which the United States emerged as a dominant world power within a new global order of bi-polar, superpower conflict. The basic validity of this formulation seems indisputable, but it has obscured an equally valid set of continuities that tie the Cold War to a longer-term history of imperial power and global affairs stretching at least as far back as the last third of the nineteenth century. Accounts of Cold War American Science have likewise viewed World War II, especially the wartime mobilization of science that culminated in the nuclear destruction of Hiroshima and Nagasaki, as the moment of rupture which brought forth a new and deadly fusion between science and state power. This emphasis on disjuncture has produced important insights about the politics of knowledge, the relationship between scientists and the state, and the centrality of scientific research to national and international histories of the Cold War, but it has generally left questions about the continuities across the divide of 1945 both unasked and unanswered. This essay attempts to open a discussion about how consideration of the pre-World War II era of empire might shed new light on the history of American science during the Cold War. In particular, the conceptualization of the period from the late nineteenth century to the end of the Cold War as a single epoch highlights the durability of connections between social knowledge and state power, as well as the close ties of American patterns of knowledge production and deployment during the Cold War to an earlier configuration of the global order built around colonialism and imperialism.[1]

1 Frederick Cooper and Randall Packard's path-breaking edited volume, *International Development and the Social Sciences: Essays on the History and Politics of Knowledge* (Berkeley, CA and Los Angeles, 1997) highlighted such long-term continuities in the history of social knowledge. Most of the extant historical research on social science, American state power, and the Cold War, however, has continued to focus exclusively on the post-World War II period.

 | DOI 10.1163/9789004264229_010

Studies of Cold War American Science have sometimes treated the social sciences as an afterthought, but social scientific inquiry was essential to the foreign policy objectives of the Cold War American State. Development policy, security studies in its myriad guises, and other state-sponsored efforts to use scientific forms of understanding to serve American Cold War objectives all drew heavily upon the social sciences and related fields to create knowledge about peoples, societies, nations, and regions. Social scientific conceptions of development, for example, concentrated upon providing the means and tools of modernization and the modern state to allied nations abroad, particularly in the 'Third World,' while fields as diverse as peasant studies, psychological warfare, and counterinsurgency sought control of the human person, by trying to dissect the psyche for the purpose of actively shaping opinions and attitudes, and by attempting to establish the physical and military conditions for security in places dominated by political and military conflict.

This history of Cold War Social Science should not be considered in chronological isolation, however, but instead tied directly to the larger historical context of the rise of the modern state and the colonial contexts that helped shape social knowledge and its deployment by the state from the nineteenth century to the post-World War II period. Cold War history, at least from the American perspective, has too often been written as if little of significance preceded the US-Soviet conflict other than World War II.[2] A growing range of studies of knowledge and imperial power, however, as well as histories of decolonization that cross the dividing line of 1945, have made it imperative to understand how the Cold War order emerged from the previous configuration of the international system of colonial empires, with its movements of knowledge between metropoles and peripheries, as well as how patterns and relationships established much earlier continued on into the post-war era.[3] In retrospect, it

2 Although I have confined my generalizations to the historiography on the United States, the chronological limitations that dominate the *Cambridge History of the Cold War* (the most recent and ambitious effort to delineate Cold War historiography from a fully global perspective) suggest that the conceptual problem is not merely one of American provincialism. See Melvyn P. Leffler and Odd Arne Westad (eds.), *The Cambridge History of the Cold War*, 3 vols. (Cambridge and New York, 2010).

3 For an introduction to studies of knowledge and imperial power, see the works cited below. Histories of anti-colonial movements tend to challenge the dominance of the cold war narrative by highlighting the middle decades of the twentieth century, and by uniting the pre- and post-World War II periods into a coherent epoch. See, for example, Penny M. von Eschen, *Race against Empire: Black Americans and Anticolonialism, 1937–1957* (Ithaca, NY and London, 1993); Jason C. Parker, *Brother's Keeper: The United States, Race, and Empire in the British Caribbean, 1937–1962* (Oxford and New York, 2008); and Nico Slate, *Colored Cosmopolitanism:*

should not be surprising that the colonial order did not disappear overnight, and that the history of social science and the state did not begin anew with the Cold War. Yet, only now are scholars starting to explore the continuities from the pre-war to the post-war periods that defined the relationship between knowledge and the state in the heyday of American power.

The discussion presented here is necessarily preliminary and based largely on secondary sources, but I hope, nonetheless, to make the case for crossing the chronological boundary point of 1945 and linking Cold War Science to the broader history of social knowledge, the state, and imperial power from the late nineteenth century onwards. My analysis begins by examining the historiography of Cold War Social Science and a misreading of the past that arose from the literature's origins in a particular post-World War II moment. The essay then moves to an account of the American South from the early twentieth century to the 1970s, the early twentieth-century discourse that described the region as a colonial space, and the back-and-forth movement of ideas and practices between the American South and the global South throughout the time period. The final section of the paper explores other areas of social scientific inquiry, including public health, demography, peasant studies, and counterinsurgency, where scholars might look for additional social scientific crossings from the colonial to the Cold War.

Looking Backward: Social Science, Social Knowledge, and State Power from the Cold War to the Nineteenth Century

Historians of Cold War Science have read the history of the social sciences in reverse, because they (or we) usually start from history of physics as their point of departure, rather than understanding the social and human sciences as 'present at the creation' of the modern state, if not outright constitutive of it. In the context of the Cold War and the early nuclear age, the historiography on science and the American Cold War State originated with ruminations about physics, the newly powerful political roles of American physicists, and the Manhattan Project generation's contributions to debates over the existential threat of weapons of mass destruction during the worst years of the US-Soviet conflict. In the era of science as the 'new priesthood,' no issue in US foreign relations seemed more pressing than the need to rein in the nuclear arms race and its threat to civilization, if not life itself. Consequently, in the immediate

---

The Shared Struggle for Freedom in the United States and India (Cambridge, MA and London, 2012).

post-World War II period, physicists—along with a whole range of newly-minted arms control experts who both invented and promoted security studies as an academic endeavor tied intimately to policymaking—entered the defense and national security establishment, which was itself newly expanded as part of the US governmental apparatus of the Cold War. Throughout the following four decades, the nuclear age and arms control served as the foremost public symbols of the fusion between modern science and state power, and historians of science did not question the primacy of physics in studies of science and politics, given both the status of the discipline within the exact sciences and its attachment to political power.[4]

Historians of science came late to the history of the social sciences and the Cold War, because they began by merely reproducing the received hierarchy of knowledge that dominated mid-twentieth century thought and made physics seem like the natural and most representative object for historical investigations of science. In the United States, contentious discussions over the scientific status of the social sciences, particularly in the context of increasingly conscious efforts by social scientists themselves to model their disciplinary practices on the natural sciences, only reinforced the impression of their fields of inquiry as sickly offshoots of the natural sciences. Although social scientific inquiry enjoyed strong connections to American politics and public policy during the period from turn-of-the-century progressivism to the New Deal of the 1930s, as well as a powerful institutional base in the Social Science Research Council and other organizations, after World War II, the social sciences seemed like poor country cousins compared to the physical sciences. As Mark Solovey has recently observed, the early post-war debate over the inclusion of the social sciences in the proposed National Science Foundation ultimately crystallized "the position of the social scientists as second-class citizens in the emerging post-war federal science system." One would have hardly known that just a decade or so earlier, economists, political scientists, sociologists, and other social scientific experts worked as key policy innovators under the New Deal, and that during the 1930s they had far greater access to political power than their counterparts in the natural sciences. Even though American

4 A range of studies from the 1950s and 1960s set the tone for the historical work that followed. Some of the early classics include Don K. Price, *Government and Science: Their Dynamic Relation in American Democracy* (New York, 1954); Robert Gilpin, *American Scientists and Nuclear Weapons Policy* (Princeton, NJ, 1962); Robert Gilpin and Christopher Wright (eds.), *Scientists and National Policymaking* (New York, 1964); Price, *The Scientific Estate* (Cambridge, MA and London, 1965); and Ralph E. Lapp, *The New Priesthood: The Scientific Elite and the Uses of Power* (New York, 1965).

social scientists maintained, and even expanded, their relationship to the state in the Cold War years, they nonetheless perceived themselves as losing methodological, financial, and political status to the natural sciences.[5]

This historical trajectory, along with the political debates over the Cold War and its ideological pressures, has meant that scholarship on the political history of Cold War Social Science has generally mirrored the existing narrative for the history of the natural sciences. The prevailing questions in the field have usually revolved around the relationship between academic knowledge and the Cold War State; the power of patronage to shape research agendas, redefine existing disciplines, and create entire new 'knowledge communities'; the militarization of social scientific inquiry; critics and sources of resistance; and the extent to which the objectives of the national security state came to dominate social scientific inquiry, as opposed to the persistence of alternative agendas that cannot be reduced to Cold War concerns.[6] A longer-term

5 Mark Solovey, *Shaky Foundations: The Politics—Patronage—Social Science Nexus in Cold War America* (New Brunswick, NJ and London, 2013), p. 21. On the significance of social knowledge in American policy making and reform movements in the decades prior to World War II, see Mary O. Furner and Barry Supple (eds.), *The State and Economic Knowledge: The British and American Experiences* (Cambridge and New York, 1990); Ellen Fitzpatrick, *Endless Crusade: Women Social Scientists and Progressive Reform* (Oxford and New York, 1990); Mary O. Furner, "Social Scientists and the State: Constructing the Knowledge Base for Public Policy, 1880–1920," and Leon Fink, "Expert Advice: Progressive Intellectuals and the Unraveling of Labor Reform, 1912–1915," both in Leon Fink, Stephen T. Leonard, and Donald M. Reid (eds.), *Intellectuals and Public Life: Between Radicalism and Reform* (Ithaca, NY and London, 1996), pp. 145–81 and pp. 182–213, respectively; Maurine W. Greenwald and Margo Anderson (eds.), *Pittsburgh Surveyed: Social Science and Social Reform in the Early Twentieth Century* (Pittsburgh, PA, 1996); Leon Fink, *Progressive Intellectuals and the Dilemmas of Democratic Commitment* (Cambridge, MA and London, 1997), chapters 1–3; Jess Gilbert and Ellen Baker, "Wisconsin Economists and New Deal Agricultural Policy: The Legacy of Progressive Professors," *Wisconsin Magazine of History* 80 (1997), 281–312; Helene Silverberg (ed.), *Gender and American Social Science: The Formative Years* (Princeton, NJ, 1998); Jess Gilbert, "Low Modernism and the Agrarian New Deal: A Different Kind of State," in Jane Adams (ed.), *Fighting for the Farm: Rural America Transformed* (Philadelphia, PA, 2003), pp. 129–46; Thomas A. Stapleford, *The Cost of Living in America: A Political History of Economic Statistics, 1880–2000* (Cambridge and New York, 2009); and Jessica Wang, "Local Knowledge, State Power, and the Science of Industrial Labor Relations: William Leiserson, David Saposs, and American Labor Economics in the Interwar Years," *Journal of the History of the Behavioral Sciences* 46 (2010), 371–93.

6 See, for example, Christopher Simpson (ed.), *Universities and Empire: Money and Politics in the Social Sciences during the Cold War* (New York, 1998); Ron Robin, *The Making of the Cold War Enemy: Culture and Politics in the Military-Industrial Complex* (Princeton, NJ, 2001); David C. Engerman, "American Knowledge and Global Power," *Diplomatic History* 31 (2007), 599–622; Engerman, *Know Your Enemy: The Rise and Fall of America's Soviet Experts* (Oxford

perspective, however, suggests the need to reframe this approach by taking Cold War studies out of the exigencies of its immediate historical moment. Rather than depicting the Cold War as an abrupt break from the past, scholars need to investigate the continuities that persisted across the historical divide of World War II. Christina Klein has highlighted the need to interrogate the standard view of the Cold War as "a unique historical era defined by the conflict between the United States and the Soviet Union," and to see the epoch instead as "a chapter in the ongoing process of globalization."[7] I would go further and emphasize the importance of seeing the Cold War within the context of the evolution of the modern state, as well as understanding it as a particular rearrangement of relations within the international system. As much as the Cold War order represented a new historical development, it was not *sui generis*, and scholars need to recognize more fully its roots in the imperialist great power system of the previous century.[8]

In particular, social science, social knowledge, and their fusion with state power long predated the rise of physics to political prominence, and the social and human sciences provided key resources for the extension of state power in both national and colonial contexts from the mid-nineteenth century onwards. Public health (which belongs as much to the social as the biomedical arena), agriculture (another field of knowledge at the border of the natural and human worlds), anthropology and ethnography, geography, political economy, demography, and other fields of social inquiry were essential to the basic project of the modern state, whether in consolidating the nation or in facilitating colonial administration.[9] The modern state's mission of making its people and

---

and New York, 2009); Joel Isaac (ed.), "The Human Sciences and Cold War America," Special issue of *Journal of the History of the Behavioral Sciences* 47 (2011), 225–321; Mark Solovey and Hamilton Cravens (eds.), *Cold War Social Science: Knowledge Production, Liberal Democracy, and Human Nature* (New York, 2012); Solovey, *Shaky Foundations* (2013); Joy Rohde, *Armed with Expertise: The Militarization of American Social Research during the Cold War* (Ithaca, NY and London, 2013).

7 Christina Klein, *Cold War Orientalism: Asia in the Middlebrow Imagination, 1945–1961* (Berkeley CA, and Los Angeles, 2003), p. 16. I thank John Krige for alerting me to Klein's observation.

8 Similarly, as Mark Mazower has argued, internationalism originated as an expression of great power imperialism, even as its universalist claims also provided powerful conceptual resources for anti-colonial movements. Mazower, *No Enchanted Palace: The End of Empire and the Ideological Origins of the United Nations* (Princeton, NJ, and Oxford, 2009).

9 On social knowledge and state power in colonial contexts, see, for example, Helen Tilley, *Africa as a Living Laboratory: Empire, Development, and the Problem of Scientific Knowledge, 1870–1950* (Chicago and London, 2011). On social knowledge and the rise of the modern state in general, consult Alain Desrosières, *The Politics of Large Numbers: A History of Statistical*

resources legible, visible, and, as a result, an object of oversight and control, depended on all manner of social data-gathering, in which censuses, maps, cadastral surveys, and other types of social knowledge provided crucial tools of governance.[10] Far from being a Cold War afterthought, social scientific forms of inquiry played critical roles in the emergence of the modern state and the development of state capacity from the nineteenth century onwards.

Scholars therefore need to consider how attention to global history over the long term, especially the ties between imperial order and US Cold War objectives, can illuminate the Cold War history of the social and human sciences in the American context. Recent research on the history of science and colonialism has underscored the centrality of colonial settings to the generation of scientific knowledge, as well as the movements of ideas and people that shaped the fusion of knowledge to the practices of state power in both periphery and metropole.[11] Here I want to connect the insights from this literature to the Cold War deployment of social knowledge by the United States in order to suggest how the dramatic post-World War II era of technical and economic assistance programs rested upon decades of developmentalist thought and action. Moreover, even as the transition from an era of empires to one of nations forced the conceptual apparatus of development to shift from civilization to modernization, and from colonial rule to the sovereign nation-state, the same discourses and problems of power and control persisted in the appeal to social science for means to effect the wholesale transformation of societies worldwide.

Reasoning, trans. Camille Naish (Cambridge, MA and London, 1998; originally published as *La politiques des grands nombres: Histoire de la raison statistique* (Paris, 1993)), especially chapters 5–6; and Theodore M. Porter, *Trust in Numbers: The Pursuit of Objectivity in Science and Public Life* (Princeton, NJ, 1995), especially chapters 6–8.

10 See James C. Scott, *Seeing Like a State: How Certain Schemes to Improve the Human Condition Have Failed* (New Haven, CT and London, 1998); and Timothy Mitchell, *Rule of Experts: Egypt, Techno-Politics, Modernity* (Berkeley and Los Angeles, 2002).

11 For multiple perspectives on this point, see, for example, Tilley, *Africa as a Living Laboratory*; Warwick Anderson, "Introduction: Postcolonial Technoscience," *Social Studies of Science* 32 (2002), 643–58; Anderson, "Pacific Crossings: Imperial Logics in United States' Public Health Programs," in Alfred W. McCoy and Francisco A. Scarano (eds.), *Colonial Crucible: Empire in the Making of the Modern American State* (Madison, WI, 2009), pp. 277–87; and Alfred W. McCoy, *Policing America's Empire: The United States, the Philippines, and the Rise of the Surveillance State* (Madison, WI, 2009).

Colonial Crossings: From the American South to the Global South

The history of public health and economic development efforts in the American South provides one entryway into this discussion, by showing the easy flow between the local and the international, along with the complex discourses of race and underdevelopment that united American perceptions of both the South and the Third World throughout much of the twentieth century. Hookworm eradication provides an appropriate starting point for following this back-and-forth dynamic in the history of American developmentalism as it moved from the era of early twentieth-century global imperialism to mid-twentieth-century superpower conflict. In October 1909, a large donation by John D. Rockefeller, Sr. led to the establishment of the Rockefeller Sanitary Commission for the Eradication of Hookworm Disease (RSC), which launched a massive public health campaign to eliminate hookworm in the American South. The effort placed the region within a global and imperial order in at least two respects. First, the emphasis on disease as a barrier to Southern economic development reflected an early twentieth-century geographical discourse about the tropics that associated the American South with the disease-ridden, benighted, and ungovernable regions of the world. As Natalie J. Ring has observed, American cultural geographers spoke about the South in the same terms that they used to describe the dangerous, insalubrious spaces of American colonial possessions, in a discourse that tied "domestic imperial tendencies with foreign imperial impulses."[12] When members of the RSC described to John D. Rockefeller the "lowered vitality of multitudes" in the South, whose poor health was "seriously affecting economic development," they invoked an image already long-established in American and European imperial ruminations about supposed tropical backwardness.[13] "Unhooking the Hookworm," a short silent film that the Rockefeller Foundation's International Health Division made in 1920 as an educational tool for audiences in the American South, also underlined the region's shared parasitical heritage with what were, for Americans, alien and exotic parts of the world. As the opening frame of the film put it: "In all these warm countries, dwells one of man's most dangerous enemies—the Hookworm." The following

---

12 Natalie J. Ring, "Mapping Regional and Imperial Geographies: Tropical Disease in the U.S. South," in McCoy and Scarano, *Colonial Crucible*, pp. 297–308, on p. 300.

13 Rockefeller Sanitary Commission board members to John D. Rockefeller, Sr., 26 October 1909, available online at http://rockefeller100.org/items/show/2162 [accessed September 2015]. Original in Rockefeller Archive Center, Sleepy Hollow, NY, Office of the Messrs. Rockefeller Records, series O, box 52, folder 544.

seconds featured an animated map that illustrated the hookworm's habitat in large swaths of Africa, South and Central America, Southeast Asia, India, and Australia, in addition to the American South. Shortly thereafter, another bit of text intoned, "Their victims are counted by the millions, from India..." The film then used visuals to complete the thought, with footage of a crowd milling on a placid street somewhere in India, followed by a cut to an illustration of typical American children in front of a simple wood frame dwelling, presumably in the rural South. Visually, the subtext seemed clear: the American South belonged to the impoverished, less civilized parts of the world, and required the benevolent intervention of public health officials if it was to enjoy the benefits of modernity and progress.[14]

Second, the institutional networks of global hookworm control efforts coalesced, at least in part, around this imperial cultural imagery of tropical misery. The RSC's hookworm eradication mission adopted strategies developed in America's colonial empire, and it formed the basis for Rockefeller-funded health initiatives around the world in subsequent decades. At the outset of the War of 1898, US army surgeon Bailey K. Ashford participated in the military occupation of Puerto Rico, now a part of America's newly acquired overseas empire, along with the Philippines, Guam, and Hawai'i. There he identified hookworm as the cause of a pernicious anemia endemic among Puerto Rican peasants and agricultural laborers, and in the early 1900s, Ashford organized a large-scale treatment and prevention program through the Puerto Rican colonial government's newly formed Anemia Commission. Wickliffe Rose, head of the RSC, learned of Ashford's work, and in May 1910 he visited Puerto Rico to study the Anemia Commission's methods, which the Rockefeller commission adopted as a model for its own fight against hookworm in the American South.[15] The RSC itself soon attracted international attention. In 1911, former Harvard University president Charles W. Eliot sensed the international possibilities early on, when he asked Rockefeller trustee and advisor Frederick W. Gates for permission to pass along materials from the Rockefeller Sanitary Commission "to health officers in India, Java, China, and Japan," with the added observation that he presumed "they are already familiar to the American medical men

---

14 International Health Division, Rockefeller Foundation, "Unhooking the Hookworm" (1920), available online at http://www.rockarch.org/feature/hookworm.php [accessed September 2015].

15 John Ettling, *The Germ of Laziness: Rockefeller Philanthropy and Public Health in the New South* (Cambridge, MA and London, 1981), pp. 29–30, 124–27; Warwick Anderson, *Colonial Pathologies: American Tropical Medicine, Race, and Hygiene in the Philippines* (Durham, NC and London, 2006), p. 195.

in the Philippines."[16] His references to British colonial India, Dutch-ruled Java, and the American-controlled Philippines, along with a China battered by territorial concessions to outside powers, suggested the extent to which imperial networks provided ready transmission lines for the modernizing aspirations of public health. These networks grew more tightly bound two years later, when the RSC's operations in the American South became the springboard for the Rockefeller Foundation's new global health initiative, the International Health Commission (IHC, subsequently renamed the International Health Board (IHB), and then the International Health Division (IHD)).[17]

As a disease with a single, identifiable cause that responded well to cheap chemical therapy and improved sanitation, hookworm promised the Rockefeller Foundation an opportunity to showcase the power of dramatic sanitary intervention.[18] Hookworm therefore occupied center stage in the IHC's early work. The commission immediately sought new sites for hookworm

16 Charles W. Eliot to Frederick T. Gates, 21 October 1911, available online at http://rockefeller100.org/items/show/2181 [accessed September 2015]. Original in Rockefeller Archive Center, Sleepy Hollow, NY, Office of the Messrs. Rockefeller Records, series O, box 52, folder 545.

17 John Farley, *To Cast Out Disease: A History of the International Health Division of the Rockefeller Foundation (1913–1951)* (Oxford and New York, 2004), pp. 2, 4.

18 The attraction of disease eradication as a way to demonstrate the power of public health, as well as a cycle of dramatic promise followed by diminished enthusiasm after complete elimination proved more difficult to achieve than anticipated, also perpetuated itself in the campaign against yellow fever in the inter-war years and in malaria eradication efforts, particularly after World War II. In the case of yellow fever, the discovery in South America of vectors other than the *Aedes aegypti* mosquito discouraged Rockefeller Foundation efforts. The development of a vaccine in the 1930s finally contained the disease as a public health threat, although declining vaccination rates could eventually result in the return of yellow fever. Marcos Cueto (ed.), *Missionaries of Science: The Rockefeller Foundation and Latin America* (Bloomington, IN and Indianapolis, IN, 1994), pp. x–xiii; J. R. McNeill, *Mosquito Empires: Ecology and War in the Greater Caribbean, 1620–1914* (Cambridge and New York, 2010), pp. 312–4. Malaria eradication proved possible in the ecologically contained island of Taiwan, but not on a continental scale in Africa and Asia. The limitations of more ambitious malaria eradication programs and the lost prospects of Cold War political advantages that would have followed from dramatic success led the United States to turn to supporting the international program for smallpox eradication instead. Liu Yi-ping and Liu Shiyung, "A Forgotten War: Malaria Eradication in Taiwan, 1905–65," in Angela Ki Che Leung and Charlotte Furth (eds.), *Health and Hygiene in Chinese East Asia: Policies and Publics in the Long Twentieth Century* (Durham, NC, 2010), pp. 183–203; Erez Manela, "A Pox on Your Narrative: Writing Disease Control into Cold War History," *Diplomatic History* 34 (2010), 299–323.

campaigns, and during its first decade of operation, it devoted more than half of its disease-fighting budget to hookworm.[19] For Rockefeller health officials, the disease offered an ideal vector for spreading the gospel of public health worldwide, by latching broader educational and sanitation projects, as well as campaigns against other infectious diseases, onto the parasitic nematode. The IHC immediately branched out into the larger sphere of global empire. Just three months after the commission's establishment, Wickliffe Rose headed to London to consult with officials in the British Colonial Office and construct a strategy for replicating the Rockefeller's hookworm eradication program throughout British colonial possessions. In the fall, Rose then embarked upon a two-month survey of health conditions in the greater Caribbean, in a regional grand tour of British colonial order in Barbados, Trinidad, British Guiana, Grenada, St. Vincent, St. Lucia, and Antigua.[20] The path from Puerto Rico to the American South now led to the British colonial world, and Rockefeller-sponsored hookworm campaigns ultimately reached more than fifty countries located on the globe's tropical midriff.[21] In Africa and Asia, most of those countries fell under direct colonial rule, while independent Latin American nations existed in uneasy relationships with the world's dominant powers, particularly the United States, with its long record of interventionism throughout the Western Hemisphere.

The end results of the Rockefeller Foundation's widely dispersed and ambitious crusade against hookworm varied widely with local political circumstances. In particular, programs generally failed at establishing long-term public health legacies in colonial contexts, but scored some impressive successes in independent nations, especially in Latin America. In the far reaches of British, Dutch, and American empire, Rockefeller officials' aspirations for transformative public health campaigns repeatedly conflicted with the strictures of tight-fisted colonial rule. British colonial administrators in British Guiana, Ceylon, India, Sri Lanka, and elsewhere, for example, eagerly accepted Rockefeller dollars in order to combat a disease that they associated with a

19 James Gillespie, "The Rockefeller Foundation, the Hookworm Campaign, and a National Health Policy in Australia, 1911–1930," in Roy MacLeod and Donald Denoon (eds.), *Health and Healing in Tropical Australia and Papua New Guinea* (Townsville, Queensland, Australia, 1991), pp. 66–7; Farley, *To Cast Out Disease*, pp. 19, 29.

20 Ettling, *The Germ of Laziness*, pp. 189–90.

21 Anne-Emanuelle Birn, "Revolution, the Scatological Way: The Rockefeller Foundation's Hookworm Campaign in 1920s Mexico," in Diego Armus (ed.), *Disease in the History of Modern Latin America: From Malaria to AIDS* (Durham, NC and London, 2003), pp. 159–82, on p. 161.

weakened labor force and diminished economic prospects in the world's tropical zones, but balked at investing their own meager budgets into the sanitary infrastructure necessary for reducing reinfection rates. Plantation owners who viewed domestic laborers as easily replaced commodities also refused to build latrines and implement other sanitary improvements.[22] In Dutch-controlled Suriname, planters and the colonial government similarly quarreled with the Rockefeller Foundation over who should fund major infrastructure projects, and when the American philanthropic organization abandoned its hookworm enterprise there in 1923, it cited Suriname's "official ingratitude."[23] Nor did the foundation necessarily fare any better in American colonial territory. In Guam, for example, the US naval government wanted Rockefeller funds for sewerage projects that it could not afford on its own, and it tried to pass on the costs of its new sanitary directives along to the island's Chamorro people, who could ill afford to buy shoes and construct concrete privies.[24] Health officials also consistently downplayed the risks of treatment, and when prospective patients sought to avoid the gag-inducing thymol and Epsom salt regimen (or, in the 1920s, ingestion of carbon tetrachloride and chenopodium oil), as well as compulsory hospitalizations or other forced treatments in some colonial contexts, authorities preferred to invoke racial stereotypes and label recalcitrants as ignorant and uncivilized, rather than acknowledge their legitimate reasons for resistance.[25] Meanwhile, hookworm disease persisted in the American South as well, contrary to the Rockefeller Foundation's declaration of victory in the mid-1920s. The campaign did, however, provide a general stimulus for public health efforts in the region, a result that gratified the Rockefeller reform impulse.[26]

By contrast, historical studies suggest that hookworm control tended to fare better in Latin America, where such efforts conformed well with independent states' nation-building and modernizing missions, as well as the objectives of

22 Farley, *To Cast Out Disease*, chapters 4 and 5; Nandini Bhattacharya, *Contagion and Enclaves: Tropical Medicine in Colonial India* (Liverpool, 2012), pp. 176–80; Soma Hewa, *Colonialism, Tropical Disease and Imperial Medicine* (Lanham, MD, 1995), chapter 3.

23 Rosemarijn Hoefte, "The Difficulty of Unhooking the Hookworm: The Rockefeller Foundation, Grace Schneiders-Howard, and Public Health Care in Suriname in the Early Twentieth Century," in Juanita De Barros, Steven Palmer, and David Wright (eds.), *Health and Medicine in the circum-Caribbean, 1800–1968* (New York and London, 2009), pp. 211–26, on pp. 216–8, quotation on p. 218.

24 Anne Perez Hattori, *Colonial Dis-Ease: US Navy Health Policies and the Chamorros of Guam, 1898–1941* (Honolulu, HI, 2004), pp. 169 and 179–80.

25 See, for example, Hattori, *Colonial Dis-Ease*, chapter 6, especially pp. 154–6 and 173–8.

26 Ettling, *The Germ of Laziness*, epilogue.

an aspiring professional class of medical and public health experts. In Brazil and Costa Rica, for example, well-established public health organizations and communities dating back to the late nineteenth century provided ready infrastructures and receptive political environments for Rockefeller-funded initiatives, although longer-term commitments in Brazil waned by the mid-1920s.[27] In Colombia, medical professionals had already identified hookworm as a source of economic stagnation in the early twentieth century, and in the 1920s, the government eagerly pursued hygienic modernity in order to attract foreign trade and investment. By the 1930s, hookworm control had evolved into a generalized sanitation and public health apparatus on the part of the Colombian state, in partnership with the Carlos Finlay Institute, a joint creation of the Rockefeller Foundation, the Pan-American Sanitary Bureau, and the Colombian government, which conducted extensive research in epidemiology and infectious disease.[28] In Mexico, hookworm posed a relatively limited threat to public health, but nonetheless, as Anne-Emanuelle Birn has written, the country's revolutionary government enthusiastically partnered with the International Health Board as a means of state-building by "medicalizing rural Mexico," while Rockefeller officials happily embraced the project in order to improve US-Mexico relations, as well as their own "public health triumphalism."[29] Although the record of American interventionism in Latin America raised suspicions that imperialist motives came with Rockefeller dollars, political cultures throughout the region that stressed state-building, nation-building, and modernization as ways of strengthening national political institutions and warding off foreign control provided local and national governments, health-oriented reformers, and societies at large with the means to make hookworm control their own and thereby earn widespread support on the ground. As Steven Palmer has observed, American power and successful promotion of public health tended towards an inverse relationship in Latin America: "the greater the influence of the United States within a country, the less successful was the public health work undertaken there by the imperial

27 Christian Brannstrom, "Polluted Soil, Polluted Souls: The Rockefeller Hookworm Eradication Campaign in São Paulo, Brazil, 1917–1926," *Historical Geography* 25 (1997), 25–45, esp. p. 31; and Steven Palmer, "Central American Encounters with Rockefeller Public Health, 1914–1921," in Gilbert M. Joseph, Catherine C. LeGrand, and Ricardo D. Salvatore (eds.), *Close Encounters of Empire: Writing the Cultural History of U.S.-Latin American Relations* (Durham, NC and London, 1998), pp. 311–32, especially p. 319.

28 Christopher Abel, "External Philanthropy and Domestic Change in Colombian Health Care: The Role of the Rockefeller Foundation, ca. 1920–1950," *Hispanic American Historical Review* 75 (1995), 339–76, esp. pp. 346, 351, 356 and 358.

29 Anne-Emanuelle Birn, "Revolution, the Scatological Way," pp. 174 and 159.

philanthropic institution."[30] Conversely, where ambitious, self-styled modernizers and medical and public health elites from host nations took the lead, Rockefeller programs had the potential to take root as politically legitimate symbols of nationalist aspiration.[31]

One should take care not to overlook the colonial overtones of such national projects. To a considerable extent, nation-building and colonial rule in the late nineteenth and early twentieth centuries constituted flip sides of the same coin of modernist developmentalism, in which state power melded with ideas about the necessity of technological, economic, and social progress as means for demonstrating different societies' civilized status, as well as the legitimacy of the state itself. For example, just as US observers considered the American South a backward region in need of enlightened national intervention, different Latin American governments viewed their own societies as underdeveloped and in need of racial uplift. Fears of racial degeneration helped to motivate public health programs in Costa Rica, for example, whereas the mission of state, nation, and the deployment of social knowledge in Mexico in the 1920s and 1930s focused on turning indigenous peoples into modern subjects who could identify with Mexican nationhood.[32] Such realities serve as a reminder that as much as the nation represented an alternative to colonial rule, both forms of governance shared the modern state's drive to lay claim over peoples in distant locales and assert direct authority over their lives by making their societies legible and controllable.[33] Notably, imperial powers employed simultaneously the same technologies of rule—e.g., mass education, public health, and the development of infrastructure—to inculcate national identities at home as well as instill allegiance to the metropole along the colonial

30 Palmer, "Central American Encounters with Rockefeller Public Health," p. 326.

31 On this point, in addition to the works cited above, see also Steven C. Williams, "Nationalism and Public Health: The Convergence of Rockefeller Foundation Technique and Brazilian Federal Authority during the Time of Yellow Fever, 1925–1930," in Cueto (ed.), *Missionaries of Science*, pp. 23–51.

32 Palmer, "Central American Encounters with Rockefeller Public Health," p. 315; Victor J. Rodriguez, "The Practical Man: John Dewey, the Idea of America, and the Making of the Modern Mexican, 1923–1934" (PhD dissertation, University of California, Los Angeles, 2009). Marcos Cueto has also noted that public health initiatives served Latin American governments' desire to attract European immigration. Cueto (ed.), *Missionaries of Science*, p. xiii.

33 On the idea of legibility as a project of the modern state, see Scott, *Seeing Like a State*. Nick Cullather, "Damming Afghanistan: Modernization in a Buffer State," *Journal of American History* 89 (2002), 512–37 provides a good example of such legibility at work.

periphery.[34] For such reasons, one needs to reckon with the complexities of specific instances and not assume too readily the dichotomy of national legitimacy and colonial corruption. For example, although colonial administrations frequently sought to duck the full expense of Rockefeller-sponsored hookworm campaigns, Jamaica stands out as a notable exception to the rule of official colonial indifference to serious investment in sanitation and public health. According to James C. Riley, in British-ruled Jamaica, the International Health Board's efforts stimulated enough popular support from Jamaicans to force local parish governments to fund latrine-building and the transformation of public sanitation on the island, which brought hookworm infection under control. Meanwhile, in Australia, settler colonization created a different dynamic, in which the British Commonwealth expressed little concern about hookworm until the early twentieth century, when the White Australia policy required public health interventions to make the tropical regions of Northern Australia medically safe for white settlement, at the expense of indigenous peoples' autonomy.[35]

Amid these complex realities of varying circumstances in diverse places, the Rockefeller Foundation's International Health Division successfully fashioned itself into the central institution for the promotion of international health prior to World War II, which placed it squarely within the flow of global imperial power. Hookworm campaigns became the launchpad for a wide variety of initiatives, many of them associated with colonial rule or with the League of Nations, itself a body in which internationalist ideals arose from imperial

34 For example, at the same time that the French government sought to ensure that colonized peoples learned about "nos ancêtres les Gaulois" and their place as subjects within a great French empire, it was simultaneously working to teach peasants in rural France to choose identification with the French nation over more local affiliations: Eugen Weber, *Peasants into Frenchmen: The Modernization of Rural France, 1870–1914* (Stanford, CA, 1976), especially chapters 12 and 18. The United States similarly employed technologies of imperial rule in both its overseas colonial empire and in a mainland where high immigration rates from southern and Eastern Europe raised anxieties about the maintenance of social order and the perpetuation of American identity. See, for example, Anderson, "Pacific Crossings"; and McCoy, *Policing America's Empire*. In addition, Pablo Navarro-Rivera has provided an important reminder that colonial education policy in Puerto Rico had its origins in nineteenth-century industrial schools for African Americans and American Indians. Pablo Navarro-Rivera, "The Imperial Enterprise and Educational Policies in Colonial Puerto Rico," in McCoy and Scarano (eds.), *Colonial Crucible*, pp. 163–74.

35 James C. Riley, *Poverty and Life Expectancy: The Jamaica Paradox* (Cambridge and New York, 2005), pp. 83–91, 104; Gillespie, "The Rockefeller Foundation, the Hookworm Campaign and a National Health Policy in Australia," p. 70.

conceptions of authority and order, as Mark Mazower has shown.[36] The full history of American philanthropy, social knowledge, and colonial order has yet to be written, but existing studies provide tantalizing evidence of the relationships at work. For example, conventional accounts assume little US involvement in sub-Saharan Africa during the inter-war years outside of the political circles of the African diaspora. Helen Tilley's account of science and development in British colonial Africa, however, is punctuated with frequent appearances by the Rockefeller Foundation, which provided funding for social science research and international public health programs far beyond the resources of what Britain's own institutional apparatus could provide.[37] Hookworm formed just one part of a much larger global history of social inquiry, health, and development in the pre-World War II period.

The point here is not to suggest that hookworm eradication and other international health endeavors were nothing more than neo-colonial projects—if anything, the range of experiences described here suggests a more expansive array of meanings at work.[38] For my purposes here, the Rockefeller Foundation's public health work highlights the fluidity of relationships and activities that tied programs in the United States to broader global and imperial mobilizations of knowledge. This interplay of developmental logics in the American South and other parts of the world continued from the 1930s on into the Cold War years. As Bruce Schulman has observed, when Franklin D. Roosevelt and his New Dealers, both Northerners and Southerners, surveyed the economic structure of the United States during the Great Depression, they quite literally saw a colonial relationship. According to Schulman, "[t]hey viewed the South as a colonial economy, a source of raw materials, cheap labor and profits for the industrial north." Texas congressman Maury Maverick made the colonial reference explicit when he compared the flow of wealth northward to the movement of "African ivory out of the Congo."[39] In 1936, sociologist Howard

---

36 Tilley, *Africa as a Living Laboratory*, chapter 4; Mazower, *No Enchanted Palace*, chapters 1–2.

37 Tilley, *Africa as a Living Laboratory*, chapter 2 (esp. p. 95) and chapter 4.

38 On this matter I agree with Frederick Cooper and Randall Packard, who have highlighted the nineteenth-century colonial origins of development discourses, and also warned against monolithic accounts of development that fail to take into account the attractions of developmentalist ideology to nationalist aspirations in the twentieth century. Cooper and Packard, "Introduction," pp. 9–13, and Cooper, "Modernizing Bureaucrats, Backward Africans, and the Development Concept," pp. 65 and 83–4, both in Cooper and Packard (eds.), *International Development and the Social Sciences*.

39 Bruce J. Schulman, *From Cotton Belt to Sunbelt: Federal Policy, Economic Development, and the Transformation of the South, 1938–1980* (Durham and London, 1994), pp. 6–7, quotations on p. 6.

Washington Odum, whom Schulman has called "the South's first modern sociologist," also described the region as "a furnisher of raw materials to the manufacturing regions, essentially colonial in its economy."[40] Consequently, New Deal programs in the South, far from aiming solely at the economic exigencies of the Great Depression, sought nothing less than the complete economic and social transformation of the region.

The Tennessee Valley Authority (TVA), which formed the centerpiece of this grand project, built a massive network of dams along Tennessee waterways that distributed services to seven Southern states, along with a broader democratic utopianism that promoted multipurpose regional development through cheap electricity, flood control, disease control (particularly in the form of malaria eradication), resource management, land reclamation, agricultural programs, social and technical aid for the region's farmers, and virtually any other improvement scheme that TVA chairman David E. Lilienthal's "dreamers with shovels" could imagine. Globally, the program came to symbolize the combined force of technological, economic, agricultural, and sociological expertise when tied to state power, and high-placed visitors from around the world flocked to the American South to see TVA's facilities and envision similar forms of technological transformation in their own countries. For social scientists, the TVA represented the opportunity of a lifetime. As David Ekbladh has put it, "For the social science community the TVA had a siren's call [...]. It was an outstanding example of the evolving integrated approach to development that utilized social sciences. Every breed of social scientist was needed to do the surveys and analysis that would define what needed to be done in the valley—from education to resettlement."[41]

With the rise of the Cold War, the United States quickly moved to export the New Deal developmental vision abroad, through a series of relationships and programs that tied the pre-war to the post-war era, as well as impoverished parts of rural America to the rest of the world. Already during World War II and in the immediate post-war years, newly dubbed international development experts, many of whom possessed long experience in colonial

---

40 Odum, quoted in ibid., p. 42; Schulman's description of Odum appears on p. 41.

41 David Ekbladh, *The Great American Mission: Modernization and the Construction of an American World Order* (Princeton, NJ, and Oxford, 2010), p. 57. For the quintessential New Deal statement about the TVA as a means of realizing democratic self-rule and social progress, see David E. Lilienthal, *TVA: Democracy on the March* (New York and London, 1944); chapters 8 and 12 in particular address the centrality of experts to Lilienthal's New Deal vision of democratic developmentalism. The phrase "dreamers with shovels" appears towards the end of the book, on p. 217.

administration, adopted the New Dealers' enthusiasm for the TVA and looked to the program as a model of modernization that could be transferred and reproduced around the globe. For international organizations such as the United Nations Educational, Scientific, and Cultural Organization (UNESCO), the United Nations Relief and Rehabilitation Administration (UNRRA), the World Health Organization (WHO), the Food and Agricultural Organization (FAO), and also the International Bank for Reconstruction and Development (IRBD, now better known as the World Bank), TVA's promise of economic growth, democratization, agricultural improvement, disease control, and general social uplift offered a package solution to poverty and global disorder in the aftermath of World War II.[42] President Harry S. Truman also eagerly supported the TVA model. His 1949 inaugural address prompted the creation of the Point Four development program, when he called upon the United States to "embark on a bold new program for making the benefits of our scientific advances and industrial progress available for the improvement and growth of underdeveloped areas."[43] That pronouncement launched an era of development projects as a major component of US foreign policy more than a decade before Rostovian modernization theory came into vogue.

Truman imagined that the spread of TVA-like systems around the world would offer the best and most powerful advertisement for this developmental mission.[44] Moreover, rather than considering dams as unwanted projects imposed by external powers, modernizing leaders in places previously under imperialist pressure or outright colonial rule eagerly welcomed dams and other Point Four-funded projects as means to consolidate their own state-building and nation-building efforts. By the late 1950s, almost two dozen TVA-like undertakings operated in six continents, including a "Water for Peace" program in Jordan, the Helmand Valley Authority in Afghanistan (later expanded to the Helmand and Arghandab Valley Authority), and the

42 Ekbladh, *The Great American Mission*, pp. 80–91.

43 Harry S. Truman, Inaugural Address, 20 January 1949, text available online at the American Presidency Project: http://www.presidency.ucsb.edu/ws/index.php?pid=13282 [accessed September 2015].

44 Ekbladh, *The Great American Mission*, pp. 98–9; Cullather, "Damming Afghanistan," p. 524. In 1944, David E. Lilienthal already imagined the spread of TVA-like systems in any place with rivers large enough, whether in China, India, Brazil, Argentina, or elsewhere. Lilienthal, *TVA: Democracy on the March*, p. 2. In the 1950s and 1960s, Lilienthal's Development and Resources Corporation became a key player in American efforts to export the TVA model. See Jason Scott Smith, "The Liberal Invention of the Multinational Corporation," in Kim Phillips-Fein and Julian E. Zelizer (eds.), *What's Good for Business: Business and American Politics since World War II* (Oxford and New York, 2012), pp. 107–22.

Khuzestan development plan in Iran. In the latter two programs, Mohammad Zahir Shah of Afghanistan and Mohammad Reza Pahlavi, the Shah of Iran, both welcomed American development aid as an adjunct to their own desires to create and project the power of the modern state in their respective nations, as part of what Bradley R. Simpson has appropriately termed "authoritarian development."[45] Both efforts failed ignominiously on technological and political grounds. In Iran, planners failed to match the dam to local soil conditions, and forced relocations into poorly constructed "model" villages alienated the very populations whose loyalties the Shah's government sought to cultivate. Meanwhile, in Afghanistan, the American hope that technology would somehow make political problems disappear collapsed in the face of the deep ethnic divisions that followed from the national government's efforts to extend its political influence by relying upon favoritism towards Pashtuns. As Nick Cullather has observed with respect to the Helmand and Arghandab Valley Authority, "the engines and dreams of modernization ran their full course, spooling out across the desert until they hit limits of physics, culture, and history."[46]

New Deal agricultural programs formed another link between developmental visions for both a colonial American South and a decolonizing global South. In the United States, land-grant universities, themselves a product of mid-nineteenth century federal policy under the Morrill Act, first combined forces with the agricultural experiment stations created under the 1887 Hatch Act, and then turned towards cooperative extension programs under the Smith-Lever Act of 1914. By the 1930s, this institutional apparatus formed the basis for the 'low modernist' interventions of New Deal programs in American agricultural regions, including the rural South.[47] Point Four and other Cold War era development programs grew directly from this nexus of agricultural institutions and state power. For example, when Oklahoma A&M (Oklahoma Agricultural and Mechanical College, renamed Oklahoma State University

45 Bradley R. Simpson, *Economists with Guns: Authoritarian Development and U.S.-Indonesian Relations, 1960–1968* (Stanford, CA, 2008), p. 3. Thomas C. Field Jr.'s recent study of Bolivia also underscores the need to address the right-wing forms of Third World nationalism that abounded during the Cold War period, rather than viewing the phenomenon purely through the lens of leftist revolution. Field, *From Development to Dictatorship: Bolivia and the Alliance for Progress in the Kennedy Era* (Ithaca, NY and London, 2014).

46 A brief discussion of the Khuzestan development effort and its dismal fate appears in Ekbladh, *The Great American Mission*, pp. 230–3. On the Helmand Valley Authority and its history, see Cullather, "Damming Afghanistan," p. 515.

47 Jess Gilbert coined the memorable term "low modernism" to describe New Deal agricultural development projects. Gilbert, "Low Modernism and the Agrarian New Deal."

in 1957) offered its services to Haile Selassie's efforts to found the Imperial Ethiopian College of Agricultural and Mechanical Arts and establish agricultural extension programs in his country, the university attempted to transplant an American system of knowledge production and dissemination to Ethiopia. Other programs to support handicraft production, promote agricultural education, improve pest control, crop production, and livestock production, provide health services, increase access to clean water, and control malaria all echoed the New Deal in the South and its fusion of diverse forms of scientific, social scientific, and technical expertise with development.[48] In Ethiopia and elsewhere in North Africa and the Middle East, ground-level programs that expanded educational opportunities or introduced new varieties of livestock also represented, as Amanda McVety once described it, "a foreign policy built upon chickens, tractors, and schools."[49] This approach represented a New Deal 'low modernism' in agriculture that accompanied the high modernism of the TVA model.

This history of the New Deal abroad should not be understood as a simple diffusion of American expertise, however. Nor should it be perceived as a sharp break from the past that transcended colonialism and erased the divide between colonizer and colonized, as much as American advocates of Cold War development programs wanted to believe otherwise.[50] Rather, New Deal-style liberal developmentalism needs to be situated within the full global context of power struggles that took place in national, imperial, and international settings. All too often, American support for nationalist aspirations

48 Amanda Kay McVety, *Enlightened Aid: U.S. Development as Foreign Policy in Ethiopia* (Oxford and New York, 2012), pp. 127–30; McVety, "Truman's Point Four Program and the Creation of America's Modern Diplomatic Vision" (PhD dissertation, University of California, Los Angeles, 2006), chapter 4. On International Voluntary Services and American-sponsored agricultural development programs in Vietnam in the 1950s and early 1960s, see Jessica Breiteneicher Elkind, "The First Casualties: American Nation Building Programs in South Vietnam, 1955–1965" (PhD dissertation, University of California, Los Angeles, 2005), chapter 5. The volunteers who staffed IVS's agricultural programs generally possessed degrees from agriculture schools, and were therefore tied to the tradition of agricultural education and extension that dated back to the nineteenth century. In addition, their localized, ground level efforts seemed reminiscent of New Deal agricultural programs.

49 McVety, "Truman's Point Four Program," p. 201.

50 On American hopes for developmentalism as an escape from an imperial world order, see Cullather, "Damming Afghanistan," pp. 513 and 523. On American officials' explicit concerns to avoid perceptions of development aid as another form of imperialism, see also McVety, "Truman's Point Four Program," pp. 175–7.

took the form of alliances with right-wing authoritarian governments, which perhaps avoided neocolonialism in an abstract theoretical sense, but still heavily implicated the United States in the perpetuation of repressive rule abroad. In addition, racism at home and its intersection with economic development was also embedded within a larger global and imperial context. Brenda Gayle Plummer's recent study of African Americans and the politics of decolonization from the 1950s to the 1970s underscores the tortured politics of race and inequality that united the Civil Rights struggle with anti-colonial and anti-imperialist movements around the world, as well as the poverty and under-development that made the American South part of the global South. Her account insists that one should not accept uncritically American officials' perception of the United States as an alternative to European imperialism, but rather, that the deeply embedded history of racial discrimination in the United States needs to be seen as part of a global history that linked racism and imperialism. That shared history meant that even as the United States pursued development for the declared goals of modernization and nation-building abroad, its own developmental efforts within US borders could not transcend colonial legacies. As Plummer provocatively suggests, in the so-called New South of the post-World War II period, "southern elites exchanged formal Jim Crow for development," but they did so by relegating African Americans to a subordinate economic position that preserved the racial hierarchy of the old, underdeveloped order even as the high-technology, defense-oriented economy of the Cold War fundamentally transformed the region.[51] Plummer also juxtaposes levels of malnourishment among African Americans in the South that rivaled conditions in the Third World with NASA's space age economy. For example, extensive African American poverty and deprivation in Brevard County, Florida, home of the Kennedy Space Center, contrasted sharply with the space program's lucrative professional opportunities for a white-dominated scientific and engineering elite. American tracking stations in South Africa, where the Apartheid government prohibited NASA from hiring black workers, completed the circuit between the American racial order and the racial legacy of European colonialism.[52] African American skepticism about the necessity of the space program, and even suspicion of its truth claims, arose from this social context of a perceived gulf between social needs and political priorities. As a 1969 editorial in *Ebony* magazine derisively put it, the moon landing

---

51 Brenda Gayle Plummer, *In Search of Power: African Americans in the Era of Decolonization, 1956–1974* (Cambridge and New York, 2013), pp. 145–6.

52 Ibid., pp. 241–5, 307–15.

constituted "one small step for 'the Man,' and probably a giant step in the wrong direction for mankind."[53]

Development Discourse from the Age of Empire to the Era of Decolonization

This account of the American South and the history of development in the twentieth century has emphasized the back-and-forth interplay between circumstances within the United States and those of colonial and early post-colonial contexts. By doing so, I have tried to erode the artificial distinction between the domestic and foreign settings where social scientific expertise operates, as well as to suggest the continuity from the pre-war history of development as a colonial project to the history of American-sponsored development projects in the Cold War. The larger history of public health, demography, peasant studies, and other forms of social inquiry also exposes a broad array of colonial crossings well beyond those that tied development in the American South to a larger global order.

Recent studies of public health, for example, have established the centrality of disease control to the mission of empire at the turn of the century. Within an international system that placed a premium on civilization as the key measure of nations' fitness for both self-rule and authority over others, standards of hygiene prominently marked any given society's claim to civilized status. Hence, in American colonial and imperial contexts, public health provided both a rationale for American interventionism and a key tool of governance. Mariola Espinosa, for example, has shown how the War of 1898 grew in part from American fears of Cuba as a purported breeding ground for the yellow fever epidemics that periodically swept through the American South. For well over a decade before the war, pro-annexationists in public health, political, and journalistic circles claimed that the Spanish colonial government's inability to control yellow fever justified American acquisition of Cuba, whether by purchase or through use of force. In that sense, sanitation provided the moral justification for empire, and states that failed to maintain sanitary standards in

53 Quoted in ibid., p. 245. African American alienation towards NASA also expressed itself through higher rates of skepticism about the reality of the moon landings than among Americans as a whole. For example, a survey of African Americans in a poor neighborhood of Washington, DC in June 1970 found that more than half of respondents believed the moon landings had never taken place. "Many Doubt Man's Landing on Moon," *Atlanta Constitution*, 15 June 1970.

their colonial holdings surrendered their right to rule.[54] Furthermore, under the US occupation of Cuba that followed the war, the military government immediately erected an extensive institutional structure for street cleaning, disinfection of public facilities, and quarantine in order to prevent and contain yellow fever outbreaks. After the US Army's Yellow Fever Commission added experimental confirmation to Cuban physician Carlos J. Finlay's finding that mosquitoes constituted the sole vector of yellow fever in Cuba, the American occupation also focused on destruction of mosquito habitats and disease eradication, and under the Platt Amendment and subsequent policies, the United States imposed continued abatement of yellow fever as a condition for Cuba's nominal independence.[55]

Cuba was just one of multiple cases in which public health rode the coattails of empire at the turn of the century. For the United States, sanitation and public health provided critical technologies of rule in the Philippines and other colonial possessions acquired in 1898, as well as in other imperial projects, such as the construction of the Panama Canal and American control over the Canal Zone, or in privatized forms of empire, as in United Fruit and its plantations in Guatemala.[56] Between the 1890s and 1920s, France established Pasteur Institutes throughout its overseas colonies, at multiple sites in Indochina and French North Africa, as well as in Dakar and Madagascar. These institutes carried out public health work and research in tropical medicine essential to colonial rule and the development of colonial possessions. Public health served similar purposes throughout the British empire.[57]

---

54 Mariola Espinosa, *Epidemic Invasions: Yellow Fever and the Limits of Cuban Independence, 1878–1930* (Chicago and London, 2009), chapter 2, esp. pp. 27–8.

55 Ibid., chapters 3–5.

56 Warwick Anderson, *Colonial Pathologies*; Anderson, "Pacific Crossings"; Paul S. Sutter, "Tropical Conquest and the Rise of the Environmental Management State: The Case of U.S. Sanitary Efforts in Panama," in McCoy and Scarano (eds.), *Colonial Crucible*, pp. 317–26; and Jason Colby, "'Banana Growing and Negro Management': Race, Labor, and Jim Crow Colonialism in Guatemala, 1884–1930," *Diplomatic History* 30 (2006), 595–621, on pp. 609–10.

57 On Pasteur Institutes and the French imperial project, see John Strachan, "The Pasteurization of Algeria?" *French History* 20:3 (2006), 260–75, on p. 263; and Anne Marie Moulin, "Patriarchal Science: The Network of the Overseas Pasteur Institutes," in Patrick Petitjean, Catherine Jami, and Anne Marie Moulin (eds.), *Science and Empires: Historical Studies about Scientific Development and European Expansion* (Dordrecht, 1992), pp. 307–22. On the convergence of medicine, sanitary science, and imperial power in the British Empire, David Arnold, *Colonizing the Body: State Medicine and Epidemic Disease in Nineteenth-Century India* (Berkeley, CA and Los Angeles, 1993) provides a place to start.

The health officials who worked as colonial administrators also moved easily between colonial, national, and international spaces. Warwick Anderson, for example, has detailed how within the American empire, the Philippines and other colonized or semi-colonized places provided training grounds for a generation of public health officials who imported colonial practices and a strict discourse of hygiene and bodily self-discipline to the United States. As part of these intra-imperial movements, Allan J. McLaughlin went from the Philippines to the prestigious position of commissioner of health in Massachusetts in 1914, while Louis Schapiro served a brief stint in Milwaukee's Health Department between sojourns in the Philippines and in Costa Rica, the latter as part of the Rockefeller project of hookworm eradication. Victor G. Heiser, chief quarantine officer in the Philippines, turned down an offer to head the New York City Department of Health, but he took it upon himself to advise Haven Emerson, who accepted the post, on the adaptation of experiences in the Philippines to urban health problems in the United States. Meanwhile, Heiser followed a path similar to Schapiro's and signed up with the Rockefeller Foundation's International Health Board, thereby establishing a career that tied colonial, national, and international concerns into a fluid network of relationships.[58]

Current scholarship has only just begun to identify the direct links between this pre-World War II history of colonial public health and the Cold War order of global health projects, but even a preliminary examination suggests the strength of these connections. For example, Kornel Chang's work on US occupation forces in Korea after World War II has revealed how American officials' view of Koreans as dirty, unhygienic, and therefore in need of sanitary tutelage as a prerequisite to self-governance directly reproduced long-familiar tropes from American colonial governance in the Philippines.[59] In another case, Randall Packard has pointed out that in 1948, Secretary of State George

58 Warwick Anderson, "Pacific Crossings." In 1915, Heiser lamented the state of sanitation and public health in the United States, as opposed to the sanitary order he had helped to establish in the Philippines. He told *Harper's Magazine*, "[o]nly in the Philippine Islands do I feel secure. You people at home don't realize that sanitary conditions there are much better than here. We in the Philippines wouldn't for a moment submit to the unsanitary conditions that exist in the United States." The reporter from *Harper's* then observed, "This is certainly a new point of view: that the United States is to learn real sanitation by way of the Philippines. But it is not only novel—it is true." Burton J. Hendrick, "An American Who Made Health Contagious," *Harper's Magazine* 132 (1 December 1915), 717–25, on p. 717.

59 Kornel S. Chang, "An Intimate Geo-Political Reversal: Koreans under U.S. Occupation and the Postwar Redemption of Japan, 1945–1948," unpublished manuscript, esp. pp. 1–3, 9–12,

C. Marshall spoke to the Fourth International Congress of Tropical Diseases and Malaria in the by then well-established vernacular of tropical disease as an impediment to regional social progress and world economic prospects. His address harkened back to early twentieth century discourses of tropical underdevelopment, by emphasizing both the human costs of disease, which left "millions weak and inefficient," as well as the economic damage that undeveloped resources caused in an interdependent global marketplace.[60] Recent studies have also become increasingly conscious of the continuity of expert personnel that accompanied the transition from colonial administration to a world of new nations and new international organizations. The ubiquitous Julian Huxley, energetic evolutionary biologist and internationalist, provides a typical example. Huxley circulated readily between the British scientific establishment, the British colonial research apparatus, and international organizations such as the League of Nations, and he moved easily from appealing to the League or the Rockefeller Foundation to support the African Research Survey in the interwar years to becoming director of UNESCO right after World War II. In the 1960s, when Huxley promoted environmental conservation efforts through the International Union for the Conservation of Nature and the World Wildlife Federation, the internationalist mission to incorporate natural African landscapes into a protected global human heritage sounded uncomfortable colonial overtones to a continent undergoing rapid decolonization and the emergence of countries eager to assert national independence. Although environmentalism offered newly founded nations novel forms of identity and economic possibility, the assertion of international interests also raised the ominous prospect of incursions upon sovereignty and the removal of newly liberated land from national control.[61]

The continuity of individuals and ideas from colonial administration to international organization also persisted in other areas of social scientific endeavor. A recent essay by Samantha Iyer has pointed to the relationship

---

and 15–25. I am grateful to Kornel Chang for permission to cite this fascinating and important work in progress.

60 Randall Packard, "Visions of Postwar Health and Development and Their Impact on Public Health Interventions in the Developing World," in Cooper and Packard (eds.), *International Development and the Social Sciences*, p. 97.

61 On Huxley and research in British colonial Africa, see Tilley, *Africa as a Living Laboratory*. On his conservation agenda after World War II, see Stephen Joseph Macekura, "Of Limits and Growth: Global Environmentalism and the Rise of 'Sustainable Development' in the Twentieth Century" (PhD dissertation, University of Virginia, 2013), chapters 1–2.

between demography and development, and "how different development concepts arose out of a dialogue between British colonial, Indian nationalist, and American thinkers."[62] Starting in the late nineteenth century, British social scientists concerned with India began to reckon with demography as they imagined how improved sanitary conditions and declining mortality might produce an untenable growth in population. Such ideas were translated into an American context by Edward A. Ross and other sociologists concerned about immigration and an influx of racial others that might swamp native-born, white Protestant, Euro-American populations. British readers such as Alexander Carr-Saunders, director of the London School of Economics and an advisor to the Commission on Higher Education in the Colonies, then incorporated Ross's ideas into British concerns about colonial management.[63] In this case, imperial crossings joined the Atlantic crossings that defined the transnational character of progressive social science in the decades around the turn of the century. Subsequently, pre-war discourses about demographic transition theory moved into post-World War II discussions about development, partly through continued trans-Atlantic exchanges and appropriations of ideas, and partly through the direct movements of experts. Arthur Lewis, for example, worked in the British Colonial Office from the late 1930s to the late 1940s, before joining Princeton's economics faculty. His work in the 1950s on underdevelopment, surplus labor, and the challenges of societies with high fertility and low mortality eventually earned him a Nobel Prize. Iyer has also pointed out that the very concept and label of the 'Third World' arose precisely from French demographer Alfred Sauvy's concerns about French Algeria and other places that combined high fertility with low mortality. Iyer's provocative account thus suggests the various paths by which colonial discourses about population became deeply embedded in Cold War distinctions between modern and traditional societies and discussions about development.[64]

---

62 Samantha Iyer, "Colonial Population and the Idea of Development," *Comparative Studies in Society and History* 55 (2013), 65–91, on p. 68.

63 Ibid., pp. 83–7.

64 Ibid., pp. 88–90. On Arthur Lewis, see also Cooper, "Modernizing Bureaucrats, Backward Africans, and the Development Concept," pp. 82–3. On the "Atlantic crossings" that defined progressivism from the 1870s to the 1940s, see Daniel T. Rodgers, *Atlantic Crossings: Social Politics in a Progressive Age* (Cambridge, MA and London, 2000). In addition, the Rockefeller Foundation became a major supporter of demographic research in the 1950s and 1960s, a development that John Sharpless once compared to the foundation's public health work in the inter-war period. Sharpless, "Population Science, Private Foundations,

Cold War peasant studies, as a field of social science that sought to explain the mindset of poor, rural peoples allegedly mired in tradition, also all too easily transferred old, racialized colonial stereotypes into post-war social science. Timothy Mitchell has ably dismantled Richard Critchfield's famous 1978 account, *Shahhat: An Egyptian*, and shown how it replicated familiar caricatures about timeless villagers and the easy violence of undeveloped peoples, not to mention entire passages from Jesuit priest Henry Habib Ayrout's 1938 study of peasant society in Egypt. Critchfield's literal (and liberal) borrowing from Ayrout directly bridged colonial-era sociology and Cold War peasant studies. In addition, Ayrout's classic work also circulated directly in American social science and policy circles, through a 1945 edition published by Yale University's cultural anthropology data-gathering project, the Human Relations Area Files, and as a later American translation by the Point Four program. In Mitchell's rendering, the worst offenses of *Shahhat: An Egyptian* lay not in the wild inaccuracies of Critchfield's account, the outright plagiarism, or the possibility that Critchfield was working for the CIA, but in the academic community's uncritical embrace of the book as an authentic and insightful account of village life in Egypt. Critchfield earned rich academic rewards for his writings on peasant life, including Ford Foundation support for *Shahhat: An Egyptian*, a subsequent fellowship from the Rockefeller Foundation, and a MacArthur 'genius' scholarship.[65] In an era still steeped in preconceptions about the irrationality of peasant adherence to tradition, it was no wonder that when James C. Scott and Samuel L. Popkin produced incisive studies from competing perspectives that argued for peasants' rationality and their savvy about the political and power relationships that shaped their lives and opportunities, the new scholarship seemed utterly innovative and remarkable.[66]

---

and Development Aid," in Cooper and Packard (eds.), *International Development and the Social Sciences*, p. 183.

65 Mitchell, *Rule of Experts*, chapter 4. Critchfield's Ford and Rockefeller support is noted in the front matter of Richard Critchfield, *Shahhat: An Egyptian* (Syracuse, NY, 1978), pp. iv, vi.

66 James C. Scott, *The Moral Economy of the Peasant: Rebellion and Subsistence in Southeast Asia* (New Haven, CT and London, 1976); Samuel L. Popkin, *The Rational Peasant: The Political Economy of Rural Society in Vietnam* (Berkeley, CA and Los Angeles, 1979). Stacy Leigh Pigg has observed that even when development experts try their best to behave with cultural sensitivity and seek genuine partnerships with impoverished rural peoples, stereotypes about traditional village life still tend to impede recognition of villagers' own sense of their modernity. Pigg, "'Found in Most Traditional Societies': Traditional Medical Practitioners between Culture and Development," in Cooper and Packard (eds.), *International Development and the Social Sciences*, pp. 266–70.

One could find similar colonial crossings in other areas of Cold War Social Science and related fields of expertise devoted to state projects of legibility and discipline. Methods of policing, surveillance, coercion, and outright repression that had been developed to manage populations in the Philippines found their way into the United States, via US military intelligence and the FBI. For example, as Alfred McCoy has noted, shortly after World War I, "senior US Army commanders applied lessons from the colonial Philippines to crush a radical miners' revolt in the West Virginia coalfields."[67] Post-World War II counterinsurgency drew its lessons both from the American empire and from European colonial experience, and American advocates of counterinsurgency frequently cited British success in suppressing a communist-led insurgency in Malaya as their model.[68] Critics, meanwhile, have viewed counterinsurgency as colonialism in disguise. David Marr, an intelligence officer in the Marines during the early 1960s who later became one of the foremost American historians of modern Vietnam, wrote in the *Pentagon Papers* that American counterinsurgency efforts in South Vietnam "bore striking resemblances to nineteenth-century French techniques going by the title of 'pacification,' or for that matter, earlier tactics used by Vietnamese monarchs to suppress peasant rebellions." American strategists' assumptions that "Vietnamese peasants worried only about where their next bowl of rice was coming from" or "found little meaning or value in political ideology" were also, Marr argued, reminiscent of French colonialism: "Needless to say, the French colonials had harbored such patronizing, racist ideas about the Vietnamese peasantry long before American counterinsurgency specialists picked them up."[69] The American Cold War exportation of authoritarian policing methods tied to the control of unruly populations also drew from colonial precedents in both American and European experience.[70] One might debate the extent to which this history

67 Alfred W. McCoy, *Policing America's Empire*, chapter 1, on p. 39.

68 On counterinsurgency and the fallacies involved in its appeal to historical experience, see David Elliott, "Parallel Wars? Can 'Lessons of Vietnam' Be Applied to Iraq?," and Marilyn Young, "Counterinsurgency, Now and Forever," both in Lloyd C. Gardner and Marilyn B. Young (eds.), *Iraq and the Lessons of Vietnam: Or, How Not to Learn from the Past* (New York, 2007), pp. 24–31 and 216–29, respectively.

69 David G. Marr, "The Rise and Fall of 'Counterinsurgency': 1961–1964," reprinted in Marvin E. Gettleman, Jane Franklin, Marilyn B. Young, and H. Bruce Franklin (eds.), *Vietnam and America: A Documented History*, revised and enlarged 2nd edition (New York, 1995), on pp. 206–07 and 211.

70 See McCoy, *Policing America's Empire*; and Jeremy Kuzmarov, "Modernizing Repression: Police Training, Political Violence, and Nation-Building in the 'American Century,'" *Diplomatic History* 33 (2009), 191–221.

of policing and counterinsurgency from colonialism to the Cold War belongs to the history of the social sciences, as opposed to military tradition, but Robert W. Komer's RAND-based study of counterinsurgency in Malaya, sponsored by the Advanced Research Projects Agency in the early 1970s, certainly suggests that interest in colonial precedents persisted within the entanglement of social scientific knowledge production and military 'pacification' programs.[71] The findings of American studies of counterinsurgency also crossed back into the American urban metropole, when Civil Rights and student protests threatened to upset the domestic political order, and RAND, the Institute for Defense Analysis, American University's Center for Research on Social Systems, and other social scientific research institutes that had previously churned out studies of counterinsurgency responded by using the very same techniques for outlining the challenges of pacification within the United States.[72] As with American military policing in the Philippines in the early twentieth century, techniques for pursuing order overseas again returned home in the 1970s.

It would be possible to continue in this vein and explore other topics, but by now the basic point should be clear: namely, that the history of social knowledge and the Cold War involved complex relationships not just with places around the world, but with the past as well. This idea is certainly not new—in his late 1950s classic, *The Tragedy of American Diplomacy*, William Appleman Williams looked to the late nineteenth century for the origins of American empire.[73] Where Williams concentrated on the pressures of capitalist expansion and its ideological apparatus, however, more recent work challenges scholars to think more carefully about the place of the United States within a global imperial history defined by flows of knowledge and practices tied to the projection of state power.

Elsewhere in this volume, John Krige has emphasized the need to keep the state front and center in historical analyses of globalization—that at least in the case of nuclear information (and presumably multiple other forms of information), states played the key role in regulating the circulation of knowl-

71 R.W. Komer, *The Malayan Emergency in Retrospect: Organization of a Successful Counterinsurgency Effort. A Report Prepared for Advanced Research Projects Agency* (Santa Monica, CA, 1972). On the wider range of Komer's pacification research at RAND, see Mai Elliott, *RAND in Southeast Asia: A History of the Vietnam War Era* (Santa Monica, CA, 2010), pp. 373–93.

72 Rohde, *Armed with Expertise*, pp. 136–41.

73 William Appleman Williams, *The Tragedy of American Diplomacy*, 50th anniversary edition (New York, 2009; first edition: 1959).

edge and technology. Here I would add that scholars also cannot separate the international system—composed not just of nation-states, but also international institutions, inter-governmental bodies, NGOs, philanthropic foundations, multilateral corporations, and other organizations—from accounts of knowledge circulation and the exchanges and borrowings that defined the interplay between social scientific knowledge and states' practices across not just space and national boundaries, but long expanses of time as well. Studies of science and the Cold War need to take seriously the full global contexts of science, just as historians of the Cold War are moving well beyond the US-Soviet framework to embrace fully the "global Cold War."[74] Doing so will not only take the historical narrative beyond the confines of the NATO countries and the Eastern Bloc, but also far outside the chronological boundaries of the Cold War itself. Only then can historians understand the ephemerality of the Cold War, and appreciate its place within the broader contours of world history, the history of the modern state, and the history of the international system itself.

---

74 Odd Arne Westad, *The Global Cold War* (Cambridge and New York, 2007).

PART 4

Scientific Hubris

CHAPTER 10

Cold War Atmospheric Sciences in the United States: From Modeling to Control

*Kristine C. Harper**

The wars of the twentieth century provided the United States' atmospheric science community with opportunities to make disciplinary advances. Money flowed more freely, while the demand for forecasting services increased, and with it the demand for a larger number of trained forecasters. As the US entered World War I, its military leaders realized that the weather and weather forecasting were going to be more important considerations than in earlier conflicts. The successful use of new warfare techniques, including poison gas, flame throwers, new types of ballistics, sound ranging, and, of course, aviation units (e.g., observation balloons and airplanes), all depended on advanced meteorological expertise. One did not want to lob gas canisters at enemies only to have the gas drift back over friendly positions. To address the deficit in meteorological expertise, both the US Army Signal Service and the US Navy trained hundreds of young men—many of whom had backgrounds in engineering—as meteorological observers and forecasters in the European theater. In addition to developing new forecasting techniques, they also invented new instruments for use in the battlefield, and exploited improved communication techniques. These enabled the communication of more accurate forecasts not only to the front line, but also to aviation units, which were not flying 'all weather' aircraft and thus benefitted from up-to-date meteorological information. Nevertheless, after WWI meteorological units were quickly disbanded, funding for weather units dried up, and US meteorology stagnated due to the US Weather Bureau's meager funding. Within ten years after the end of the war, the Great Depression further dampened the Weather Bureau's ability to advance meteorological forecasting and studies in the United States. At the

* The author gratefully acknowledges financial support from the American Meteorological Society's Graduate Fellowship in the History of Science, the Dibner Institute for the History of Science and Technology, the National Endowment for the Humanities (FB-53252-07), and the Tanner Humanities Center, University of Utah. I also appreciate the helpful comments received by my fellow participants in the Cold War Science conference held at the Lorentz Center.

 | DOI 10.1163/9789004264229_011

same time, academic meteorology continued to take a back seat to other physical and earth sciences.[1]

However, meteorology's prospects changed rapidly with the onset of World War II. After President Franklin D. Roosevelt announced that the US would build over 60,000 aircraft to support the war effort, it soon became evident that the several hundred meteorologists working for the US Weather Bureau would not be sufficient to provide the necessary operational support to the war efforts. Within three years, more than 7,000 men were trained as meteorologists, and an additional 10,000 as observers and meteorological technicians. They were sent across the world to new weather stations in the tropics and in higher latitudes as the theater of operations moved throughout the war. Climatological data were gathered, weather map collections were assembled, and forecasting thumb rules were developed and sent out to field locations to aid the forecasters, many of whom had never considered meteorology as a career.[2]

Although the majority of these new meteorologists left the discipline and moved into other professions after the war, those newly-minted meteorologists who were well-versed in mathematics and physics, and who remained active in the field, would change the nature of US meteorology during the remainder of the twentieth century. They achieved this change, firstly, by developing numerical weather prediction models, which would later lead to the development of climate prediction models. Secondly, they developed weather modification techniques, often referred to as weather control. Compared with the meager funding available for meteorology before World War II, funding for those branches of science that strengthened national security, including meteorological research and weather forecasting, remained strong during the Cold War, although it never approached that received by physicists. While the US armed forces were reduced in numbers after World War II, the US maintained a substantial standing army, navy, and air force during the Cold War. Those men, and their increasingly sophisticated weaponry, required more accurate

1 For a detailed discussion of military meteorology during World War I, see Charles C. Bates and John F. Fuller, *America's Weather Warriors: 1814–1985* (College Station, TX, 1986). See also Kristine C. Harper, "Meteorology's Struggle for Professional Recognition in the USA (1900–1950)," *Annals of Science* 63 (2006), 179–99; and Kristine C. Harper, *Weather by the Numbers: The Genesis of Modern Meteorology* (Cambridge, MA, 2008), chapters 1 and 2.

2 For details on the training program, see William A. Koelsch, "From Geo- to Physical Science: Meteorology and the American University, 1919–1945," in James Rodger Fleming (ed.), *Historical Essays on Meteorology 1919–1995: The Diamond Anniversary History Volume of the American Meteorological Society* (Boston, MA, 1996), pp. 511–540; Harper, *Weather by the Numbers*, chapter 3.

meteorological forecasts, regardless of whether they operated on land, under or on the surface of the oceans, or in the air at higher altitudes. Was meteorology, then, a 'Cold War science'? Or was it a science that happened to benefit from the Cold War, just as it had benefitted from earlier armed conflicts? In this article, I will argue that both statements are correct.

Modeling the Atmosphere: Benefitting from Wars, Hot and Cold

Meteorologists around the world had struggled to develop viable atmospheric theories before World War II. Vilhelm Bjerknes and his acolytes at the so-called 'Bergen School' in Norway had developed a theoretically rigorous way of defining atmospheric motion during and just after World War I. However, the non-linear 'primitive equations', which included the equations of motion and relevant hydrodynamic and thermodynamic equations, took months to solve and were not suited for the production of, for example, a 24- or 36-hour forecast for Norwegian farmers and fishermen. Therefore, Bjerknes and his students turned to a graphical method that relied on a dense network of weather observation stations that reported basic atmospheric data (e.g., temperature and air pressure) four times a day. When using these data to determine the presence of different air masses, their analysis technique allowed them to identify the boundaries between air masses—called 'fronts'—and thus to predict the weather that accompanies an air mass change. By the early 1920s, Bjerknes and his students were sharing their techniques with meteorologists across Europe and North America.[3]

While most European weather services eagerly adopted Bjerknes's techniques, the US Weather Bureau did not. Why were the European weather services more receptive? It was much easier for smaller nations to develop the network of observation stations required for implementing the method. Moreover, by sharing weather data with neighboring nations, national weather services were also able to determine weather patterns and changes, and thus to advise their inhabitants on the weather. However, in the United States—a country defined by well-populated coasts and a sparsely-populated mid-section—it was impossible to establish a sufficient number of observation stations to render air mass analysis a viable technique. An additional impeding factor was that, unlike most of the European weather services, the US Weather Bureau (USWB) employed only a small number of college-educated

3 For details, see Robert Marc Friedman, *Appropriating the Weather: Vilhelm Bjerknes and the Construction of Modern Meteorology* (Ithaca, NY, 1989).

meteorologists. Many USWB employees had started as observers, learning to read signs of changing weather conditions in the sky on the job. Observers who remained at one station for many years would sometimes be promoted to the position of meteorologist-in-charge, without having obtained any formal scientific training. Consequently, the Weather Bureau made forecasts based on rules-of-thumb, and often resisted new scientific ideas. Indeed, the Weather Bureau showed no inclination to professionalize its meteorological services until several high-profile incidents involving the bureau came to the attention of the Roosevelt administration in the early 1930s, which led to an investigation undertaken by a specially assembled team of scientists under the auspices of the National Academy of Sciences. The advisory board's recommendations—including the hiring of university-trained meteorologists and institutionalizing scientific training for its employees—did not take place until the mid-1930s.[4] A few years later, however, World War II would lead to a complete restructuring of weather services in the United States.

The U.S. military demanded that young men chosen for a crash-course in meteorology had to have either degrees in physics or mathematics, or at least a year of calculus and a year of calculus-based physics to enter the program. Consequently, the new meteorologists who emerged saw the atmosphere in a fundamentally different way than had the Weather Bureau forecasters. These new military officers saw atmospheric conditions as something that could be determined by using physical principles and mathematical calculations. Once deployed to their duty stations, they had to be ready to determine the state of the atmosphere for the operations at hand. They did not have two or three years to watch weather patterns change with the season. And so, they did the best they could with their knowledge of physics and the handful of surface and upper air observations that were available to them. They quizzed returning pilots about flight conditions once they landed, and then passed those on to the pilots preparing to take off. Once the war was over, the academically inclined amongst them returned to universities to obtain their doctorates in meteorology, determined to find a better way to understand atmospheric processes. But unlike their brethren from earlier decades, because of the war they would have more observational data at their disposal and they would have access to a new technology developed as part of the war effort: the electronic digital computer.

Within six months of VJ-Day (14 August 1945), the US Weather Bureau was actively seeking to launch numerical weather prediction techniques. Consistently short of money for its basic operations and facing increased

---

4 See Harper, *Weather by the Numbers*, chapter 1.

demand for weather services from both government and private sectors, Weather Bureau Chief Francis W. Reichelderfer was always looking for new methods that would allow his relatively small band of meteorologists to provide accurate forecasts for the nation. Reichelderfer, a former U.S. Navy meteorologist who was tapped to lead the Weather Bureau in 1938, had been in close contact with scientists at the Radio Corporation of America (RCA) laboratory in Princeton, New Jersey, who were developing a variety of meteorological instruments under government contracts. Among those scientists was physicist Vladimir Zworykin, best known for his invention of the scanning television camera. He was convinced that it was possible to use the new electronic digital computers that had come out of the war effort to not only forecast the weather, but to control it as well. While Reichelderfer thought that the idea of controlling the weather was more than a little farfetched considering the problems meteorologists had with making short-term forecasts, he was much enamored of the idea of using electronic digital computers to objectively analyze atmospheric conditions and provide forecast maps. If the latter were possible, then the time and effort devoted to subjectively analyzing surface and upper air observations to make weather maps that could then be 'moved forward' in time to create prognostic weather maps could be redirected toward making tailored narrative forecasts for weather offices' areas of responsibility. Weather office staff members would no longer have to plot meteorological observations by hand—a tedious and relatively slow process even for the most skilled plotters—hand analyze these data as they drew a weather map covering the continent, and then use wind direction and speed observations to determine how fast or slow high and low pressure systems with accompanying frontal systems were moving from West to East. If these basic tasks could be automated and made less prone to human interpretation, the result should be better forecasts delivered in less time.[5]

To this end, Reichelderfer invited Zworykin and his colleague, the world-renowned mathematician John von Neumann, to Weather Bureau headquarters in Washington, D.C., in January 1946. Their confidential meeting was dedicated to pursuing the possibility of numerical weather prediction, and involved meteorologists from the Weather Bureau, Navy, and Army, and scientists from the National Bureau of Standards. Von Neumann was to design and build the computer upon which the proposed atmospheric models would run, and both Zworykin and von Neumann promoted the possibilities and

5 For details, see Harper, *Weather by the Numbers*, chapter 4. See also, for example, Paul N. Edwards, *A Vast Machine: Computer Models, Climate Data, and the Politics of Global Warming* (Cambridge, MA, 2010).

advantages of short- and long-term weather forecasts, as well as the ultimate solution for weather problems of all sorts: weather control. Based on information apparently leaked by navy personnel who had attended the meeting, the *New York Times* reported the next day that military leaders were especially enthusiastic about the possibility of controlling the weather.[6] It might have been more difficult to gain the interest of military leaders in faster and more accurate weather forecasts, but if numerical weather prediction techniques could eventually lead to weather control, then funding would be more likely, a fact that the anonymous sources about this "confidential meeting" knew well.

In June 1946, the Office of Naval Research signed a contract that would provide funding for both von Neumann's computer and the Meteorology Project that would develop the atmospheric prediction models to be run on it.[7] While the initial projection for having a model running on the new computer was too optimistic—two years was a ridiculously short time to get a computer design off the drawing board and built, and the models were not yet on the drawing board at all—by summer 1946, the Meteorology Project was underway with a handful of newly minted mathematics-centric meteorologists, none of whom had ever spent time making an operational forecast. However, within a year, they were joined by mathematician-turned-meteorologist Jule Charney, who had spent the previous year working with meteorologists in Norway. Unlike American meteorologists, who were either academics involved solely with research or forecasters who worked for the U.S. Weather Bureau, Scandinavian meteorologists, especially Norwegian and Swedish meteorologists, had expertise and experience in both areas. It was with these meteorologists that Charney was able to refine his ideas for simplifying the so-called 'primitive equations' of the atmosphere, so that they could be solved on von Neumann's new computer. Indeed, with Charney's most basic model, just one variable was required: surface pressure. Additionally, by proposing simplifying assumptions about atmospheric characteristics (which are beyond the scope of this discussion), Charney reduced the various dynamical equations of the atmosphere to a single partial differential equation. When his results showed signs of success, the model was coded to run on the ENIAC computer at the US Army's Aberdeen Proving Grounds in Maryland. ENIAC, originally developed to solve ballistics problems during the war, had not been completed until just after the war ended, and was now available for the exploration of other kinds of problems.

6 Sidney Shalett, "Electronics to Aid Weather Figuring," *New York Times*, 11 January 1946.

7 Justification memorandum, PD #EN1–22/00028, The Institute for Advanced Study, 6 June 1946, Library of Congress, John von Neumann collection, 15/6.

In March 1950, the Meteorology Project team members headed to Aberdeen to try out their model. The machine occupied a very large room, yet had a smaller memory capacity than today's mobile phones, and its many vacuum tubes had a high failure rate. The team took 33 days to produce two 12-hour and four 24-hour forecasts. The time spent in this enterprise was unacceptable for operational weather forecasting, but despite the model's simplicity its output looked very much like a realistic meteorological chart. The Meteorology Project team, enhanced by visiting members of the "Scandinavian Tag-team" (Ragnar Fjørtoft, Arnt Eliassen, and Bert Bolin), continued to refine the models by adding higher altitude levels and increasing their complexity. By 1952, when von Neumann's new computer was operational, the models had been coded and were ready to run. The results were less than stellar—indeed, an experienced meteorologist could have produced a better prognostic weather map in less time. Nevertheless, the Weather Bureau and the military services reported the project to be a success, and made plans to establish a Joint Numerical Weather Prediction Unit (JNWPU), which was installed in 1955.[8]

After the project's start in 1946, the military services continued to fund it, while the Weather Bureau provided the staff necessary to produce hand-drawn maps which were to serve as points of comparison for the models. Once the JNWPU became fully operational, it provided computer-generated maps to Weather Bureau and military weather offices for their use. Often the forecasters on the receiving end wadded them up in disgust, recognizing that their subjective, hand-drawn maps were a superior product. But they also reported back to the JNWPU, explaining the problems with the model-derived maps and thus providing important feedback that allowed the modelers to continually revise the models. Indeed, the decision by Weather Bureau and military leaders to force their meteorologists to use the new computer-based prognostic maps as they made their operational forecasts was critical to the further development of numerical weather prediction models. Without the routine feedback from forecasters working in a variety of settings, the modelers would have been faced with models that were internally consistent, but still might have had little to do with the reality of atmospheric events.[9]

Early models were geographically and temporally restricted to relatively small areas of the Earth for relatively short periods of time. Consequently, US models focused on the continental United States because data were readily available, and the users were located there. Models developed by Carl-Gustav Rossby's team at the University of Stockholm were similarly regional, cover-

8 See Harper, *Weather by the Numbers*, chapter 6.

9 See Harper, *Weather by the Numbers*, chapter 7.

ing Northern Europe because the data were available and met the needs of their forecasters. However, meteorologists realized that numerical weather prediction would not be universally successful or useful until global models were available. Initially, the aim was to develop models that covered the entire Northern Hemisphere because dense data networks were available. Yet, these dense data networks only covered land, which still left large expanses of water for which there would never be more than a small number of ship reports arriving throughout the day. Somehow, the models had to be able to 'fill in the blanks' as they moved, for instance, storm systems on their tracks from West to East. The Southern Hemisphere presented a more difficult problem. Despite all of the new weather stations that had been established during World War II, the Southern Hemisphere had just a fraction of the weather observations that were available in the Northern Hemisphere. The Southern Hemisphere was also dominated by oceans, so its models had to be developed to account for weather over large expanses of water. To bring both types of models together into one global model, meteorologists had to gain sufficient knowledge of the general circulation of the atmosphere to account for air movement in equatorial regions, for which there was a paucity of data.[10]

The efforts of the Joint Numerical Weather Prediction Unit only lasted a few years until inter-weather service rivalries led the Weather Bureau, Navy, and Air Force to pull out and form their own numerical weather prediction units in 1961. The Weather Bureau needed to concentrate its modeling efforts on the Continental US and near-shore waters, while the Navy needed models that would provide essential support to ships at sea, and the Air Force needed models that would be effective for their high-flying aircraft and rockets. By the early 1960s, these different models diverged as the military services focused on refining models that fulfilled their operational needs anywhere in the world, and the Weather Bureau focused on domestic weather.[11]

The United States was not the only nation focusing on atmospheric models. By the late 1950s, modeling groups in Sweden, Japan, the United Kingdom, Germany, and the USSR were all producing rudimentary computer-based forecasts in an effort to increase the complexity and accuracy of their models, alongside and in reaction to the increasing computing power that was made available with each generation of electronic computers.

Efforts to determine Earth's global atmospheric circulation patterns were not directly related to numerical weather prediction: as soon as the models produced output that aided operational forecasters they had fulfilled their

10 See Harper, *Weather by the Numbers*, chapter 8.

11 Ibid.

task. But meteorologists needed to understand why, how, and where air moved throughout the atmosphere if they were going to make real advances in theory development. In the mid-1950s, computers could barely handle regional models, much less hemispheric models that included the entire atmosphere (including altitudes above 30,000 meters). However, meteorologist Norman A. Phillips, who had worked with the Meteorology Project in Princeton and with Rossby's modeling group in Stockholm, attempted to create a general circulation model that would cover an area of approximately 10,000 kilometers by 6,000 kilometers. Due to computing constraints, his atmosphere was much simpler than the 'real' atmosphere: his 'world' included no mountains, land, water, or frozen regions, and he designated its initial atmospheric conditions before he started running the model. After running his model on the computer for twelve hours, Phillips was able to produce weather maps that extended 31 days into the future, much longer than typical weather maps of the time. Although not perfect, the resulting series of maps showed atmospheric disturbances (storms) being created, moving along with the airflow, decaying, and then ultimately disappearing from the maps as they died.[12] This early success meant meteorologists could eventually expect to create predictions that would extend years into the future. As with numerical weather prediction models, computing power remained the primary limiting factor.

While weather models had come under the purview of the military services, general circulation modeling (and hence climate modeling) in the United States was centered at the Geophysical Fluid Dynamics Laboratory (GFDL), which operated under the auspices of the US Weather Bureau. The GFDL meteorologists were interested in uncovering influences on atmospheric behavior caused by changes in the chemical composition of the atmosphere, the transfer of radiation energy through the atmosphere, and the moisture content of the atmosphere. Since it was impossible to conduct experiments in the atmosphere, they worked with computer simulations; they first changed the variables associated with the aforementioned factors and then ran their models to see what would happen.

Their early experimental models produced less-than-adequate results due to restrictions caused by computer capacity: results for just two atmospheric layers could be calculated at a time. These early researchers assumed that the Earth was one smooth land surface because coding mountains, oceans, and their attendant influences on the atmosphere would have overwhelmed computing capacity. But by the early 1960s, meteorologists were able to introduce

12 Norman A. Phillips, "The General Circulation of the Atmosphere: A Numerical Experiment," *Quarterly Journal of the Royal Meteorological Society* 82 (1956): 123–164.

multiple layers and heat-related processes (e.g., changes in radiation and condensation over time) to atmospheric models, and later they added the effects of greenhouse gases such as carbon dioxide and water vapor. While modelers experimented with these factors, they were also able to investigate how air masses of different temperatures became distributed over time, which was crucial to establishing which factors most affected the climate.[13] And although military services may not be best-known for their interest in climate modeling (while their investment in weather forecasting is well-documented), military leaders showed much interest in the geopolitical changes that could be caused by a changing climate. As early as the late 1940s, Pentagon officials were concerned that shrinking Arctic ice might allow Soviet ships to move closer to North America, thus affecting strategic considerations during the Cold War. Even before the impact of the IPCC reports in the late twentieth century, military planners tried to assess how changes in temperature and precipitation (and hence the availability of fresh water) might contribute to the destabilization of different parts of the world.[14] Climate models provided them with a tool for this purpose.

Advances in meteorology and climatology, as they related to prediction and decision making by military planners, were influenced by the Cold War because the US military did not stand down in the same way that it had in the years between World Wars I and II. Not only were military planners concerned about tensions with the USSR, as the United States became embroiled in wars in Korea and Vietnam, and in smaller proxy skirmishes in Latin America, it was incumbent upon the military weather services to insure that they could provide operational forecasts in support of a variety of weapons systems. But, no matter whether the forecasts were for military users, or civilian users, such as farmers, transportation companies, and other businesses and industries, the science was the same. Theoretical developments in the atmospheric sciences benefited everyone, and thus they were enhanced by Cold War funding without being fundamentally changed by it.

---

13 Harper, *Weather by the Numbers*, chapter 8.

14 For more details on military interest in environmental change, see Ronald E. Doel, "What is the Place of the Physical Environment Sciences in Environmental History?" *Revue d'histoire moderne et contemporaine* 56 (2009), 137–64.

Weather Control: The Perfect Cold War Tool

In contrast to the atmospheric sciences, weather control was 'Cold War Science', not merely science carried out during the Cold War. What distinguished weather control from weather and climate prediction? As the development of numerical weather prediction took place, other scientists (but rarely meteorologists) set out to make the ancient dream of controlling the weather come true. Such endeavors were not new in US history. The first US government-funded experiments in controlling the weather had taken place in Texas in 1891, as massive amounts of explosives rocked the sky in attempts to 'shock' rain out of clouds, thus bringing water to grass lands suitable only for grazing cattle. In the 1920s, the US Army funded plans to spray electrified sand into clouds. The charge on the sand grains was supposed to gather cloud droplets of an opposite charge until these were sufficiently heavy to precipitate. Neither attempt to generate rain was successful.[15]

As outlined above, Vladimir Zworykin and John von Neumann intended not only to make weather predictions for both the short and long term, but also to control the weather. They thought that modifying atmospheric variables in computer models (and thus creating, in today's terms, 'virtual weather') would allow them to modify those same variables, e.g., moisture or temperature, in the physical atmosphere and obtain the desired weather whenever and wherever it was wanted or needed. In other words, they would change the variables in the computer until the model produced a weather map that looked like the desired weather and then go outside and change those same variables in the local atmosphere. Although completely impractical, this basic idea underlay their concept of how to control the weather around the world.

Meanwhile, at the General Electric Research Laboratory in Schenectady, New York, Irving Langmuir (the Nobel Prize-winning physical chemist) and his team of researchers had concluded from their wartime work on cloaking smokes and aircraft icing that clouds did not precipitate if they were lacking in 'ice nuclei'. These were minute ice crystals that would preferentially attract cloud droplets until they became heavy enough to fall to earth in the form of snowflakes or raindrops. To test their ideas, they installed a General Electric freezer in the laboratory and lined it with black velvet, which would improve the visibility of micron-sized crystals. They then exhaled into the freezer to

15 Details on early attempts, both governmental and private, to make rain are found in Clark C. Spence, *The Rainmakers: American "Pluviculture" to World War II* (Lincoln, NE, 1980). On the use of electrified sand, see, for example, "Sand-Blasting the Clouds for Man-Made Weather," *Scientific American*, 1 April 1923, 224.

produce a 'cloud', and subsequently tested a variety of powdered chemicals, as artificial nuclei, by sprinkling them into the freezer, one at a time. During an experiment in July 1946, and thus at the same time as the Meteorology Project was beginning in Princeton, New Jersey, research assistant Vincent Schaefer became frustrated by his lack of success in inducing precipitation within the freezer. Convinced that the freezer needed to be cooled to a lower temperature, he dropped in a block of dry ice. Thousands of ice crystals fluttered to the bottom of the freezer. Encouraged by this unexpected result, Schaefer shaved off increasingly small pieces of dry ice and repeated the procedure. Each time the dry ice triggered a miniature snowstorm. Langmuir took on the task of developing the theoretical underpinnings of Schaefer's results, while Schaefer continued to run experiments in the freezer.[16] By November 1946, they were ready for a field test. Langmuir, armed with binoculars, remained at Schenectady Airport to watch, while Schaefer, carrying a 2-kilogram packet of pulverized dry ice, climbed into a small aircraft with the pilot. When they reached a suitable stratus cloud with a temperature of ca. –20°C/–4°F, at an altitude of about 4,300 meters, Schaefer opened a cockpit window and scattered the dry ice along a five kilometer-long track. Back at the airport, Langmuir saw snow fall from the cloud. A few minutes later, the flat, gray stratus cloud developed the white, cauliflower-like growths of a cumulus cloud that extended upward for some 150 meters before disappearing into a "veil of snow". Five minutes later, the entire cloud turned to snow, which fell 600 meters toward the ground before evaporating.[17] The two researchers had successfully triggered precipitation from a non-precipitating cloud by introducing dry ice 'seeds'. More experiments followed, and within a few short months Langmuir had successfully obtained five years of military funding to test the efficacy of 'cloud seeding'. The Army Signal Corps and the Office of Naval Research provided the money, and the Air Force provided aircraft and personnel.[18] Project Cirrus was born.

In addition to testing dry ice seeds, Project Cirrus also investigated silver iodide seeds, whose structure mimicked that of ice nuclei. Discovered by General Electric atmospheric scientist Bernard Vonnegut, the technique involved burning charcoal briquettes, which had been soaked in a silver

16 Irving Langmuir et al., *Meteorological Research: 1 March–1 June 1947* (Schenectady, NY, 1947).

17 E.F. Black to George F. Doriot, 1 May 1947, National Archive and Records Administration II, College Park, Maryland, Records of the Research and Development Board, Record Group 330 (hereafter NARA RDB), box 459, folder 9.

18 "Army and GE to 'Make Weather,'" *New York Times*, 14 March 1947, 25. Project Cirrus would ultimately cost over $790,000, or approximately $6.8 million in 2013 dollars.

iodide-based solution. Once the soaked briquettes dried out, they were placed in metal tubes, and then burned with a propane-fueled flame, much like a gas-fired barbeque. The resulting smoke carried approximately 100 trillion silver iodide nuclei (seeds) into the atmosphere per second. Aircraft versions of these 'seeding generators' delivered seeds directly to clouds in an effort to poke holes in them and thus to allow planes to pass through without concerns about icing. Ground-based generators were used in windward mountain areas, where upslope winds carried them aloft into the clouds piled up against the mountainside, and induced precipitation.[19] Silver iodide seeds were less expensive to produce and easier to handle than dry ice. Additionally, they had the advantage of not evaporating into the dry air, and were thus able to stay airborne for a longer period of time, allowing them more opportunity to make contact with clouds.[20]

In the late 1940s and early 1950s, several US government agencies were interested in weather control. Of them, the military services also had the funds necessary to sponsor related scientific and technological research and development. While the amount of money devoted to the development of nuclear weapons and rocket delivery systems dwarfed the amount allocated to weather control research, the latter promised a number of advantages for the Cold War standoff. It became clear after the first two atomic bombs were dropped during World War II that nuclear weapons caused not just extensive damage with the initial strike, and through residual radiation which contaminated food and water supplies, but they also severely harmed those who had not been killed outright. Furthermore, since radiation could be tracked, there would always be clear evidence to establish the use of nuclear weapons by the military. Weather control, on the other hand, offered the opportunity of a traceless weapon, and thus possible deniability. Who would be able to prove that weather control techniques had been used? Even the heaviest rain or snow would likely fall within climatological extremes. Similarly, if a nation could induce a drought in another nation, and thereby deprive both the civilian population and its military personnel of access to sufficient food and water, how could anyone prove it?

Military planners were most interested in the ability to determine in advance the location and timing of precipitation. Other possible uses of weather control included the temporary dissipation of solid cloud layers to allow pilots to locate targets for bombing or airfields for landing, and the reduction of flight hazards, e.g., icing and static that could lead to lightning strikes or St. Elmo's

19 E.F. Black to George F. Doriot, 1 May 1947, NARA RDB, box 469, folder 9.

20 Langmuir et al., "Meteorological Research."

Fire (an electrical discharge that may occur when the environment's electric field is high, and may destroy expensive aircraft radar domes). None of these techniques were certain to work, but the Secretary of War held that possible military uses were sufficiently sensitive to require his complete control of press coverage of any discussions about weather control. The Army and Navy classified any practical applications of cloud seeding.[21]

Military officials had numerous ideas for exploiting weather control. They considered hindering enemy force movements by snow, inducing precipitation over allies' agricultural areas to insure sufficient moisture levels, and diverting precipitation and exhausting clouds' water content to bolster strategic aims.[22] The Army pondered clearing clouds to aid photoreconnaissance, foil the use of cloud cover by enemy defenders, and open 'holes' near drop zones to aid parachuting airborne forces. It also anticipated inflicting excessive inclement weather on enemy staging and resting areas as a way to decrease the morale of enemy troops, regulating snow and rain during peacetime training exercises for the benefit of its own troops, damaging enemy harvests and thereby destroying their will (or ability) to continue the conflict, and using snow to expose camouflaged emplacements and reveal signs of enemy activity on supply routes.[23] Indeed, the possible uses of weather control against an enemy and for the benefit of allies are restricted only by one's imagination. Battlefields could be targeted with heavy rains to bog down enemy tanks and heavy artillery, and heavy snows in mountain passes could keep enemy forces from advancing. And for those nations that were not yet committed to the USSR, weather control could also be used as a diplomatic tool to win friends by ensuring adequate amounts of fresh water needed for agriculture, industry, and domestic use. No matter what its purpose, there would be no way to trace a nation's involvement in the resulting weather.

Practical considerations aside, Project Cirrus also had a meteorological purpose: to increase understanding of the physics and chemistry behind the formation of hydrometeors (rain, snow, hail, drizzle, etc.). Researchers examined cloud microstructure, including cloud particles' water content, the size-based distribution of particles, and the vertical growth rate of clouds about which very little was known. During their initial experiments, researchers attempting to modify clouds with dry ice and other artificial nuclei tried to produce rain

21 Carl-Gustav Rossby to the Committee on Geophysical Sciences (Research and Development Board), 7 April 1947 [Confidential], NARA RDB, box 469, folder 9.

22 L.T. Morse to Executive Council (RDB), 11 August 1947, NARA RDB, box 459, folder 9.

23 Frances L. Whedon to Helmut Landsberg, 20 August 1946 [Confidential], NARA RDB, box 459, folder 9.

or snow, differentiating between the treated and untreated clouds by using smoke bombs.[24] Indeed, a major problem researchers had to overcome was how to determine whether their actions had had any effect. Did clouds precipitate because of the seeding? Or would they have precipitated in any case?

And that was the conundrum. If the clouds would have precipitated anyway, was cloud seeding really a viable undertaking? The Weather Bureau, which was running its own Cloud Physics research program, said, "No." Langmuir, who had never encountered a super-cooled (temperature below freezing) cloud that could not be modified to precipitate, said, "Yes." By 1948, the military services urgently needed to know if they could effectively exploit weather modification techniques for operational purposes. An ad hoc committee established to evaluate the research results for the Joint Chiefs of Staff determined that dry ice turned supercooled fogs into very fine ice particles, but they tended to remain suspended in the air, thus reducing visibility more than the water particles that they replaced—not an especially helpful outcome.[25] Results of seeding runs on stratus cloud decks had ranged from no observable results to holes cut through the entire cloud. After evaluating the evidence, the committee members concluded that it was too early to apply seeding techniques to military operations, but that the military services should consider seeding fog in an attempt to disperse it by using ground or airborne seeding equipment, and advising the evaluation committee of the results.[26]

While the ad-hoc committee continued its work, the Pentagon's Research and Development Board's Panel on the Atmosphere was also investigating military uses of cloud seeding. In their 1949 Technological Estimate, panel members reported that clearing airfields of fog and military areas of clouds, increasing visibility in clouds, and eliminating aircraft icing hazards were all important military applications. Based on results from the Weather Bureau's Cloud Physics Project and Langmuir's Project Cirrus, they predicted the future effectiveness of cloud seeding as follows:

---

24 Earl G. Droessler to Helmut Landsberg, 21 April 1947, NARA RDB, box 469, folder 9.

25 Francis Reichelderfer to Ross Gunn, Delbert Little, and Harry Wexler, 17 November 1948, National Archive and Records Administration II, College Park, Maryland, Records of the Weather Bureau, Record Group 27 (hereafter NARA RG 27), entry 124, 10.

26 Cloud Physics Ad Hoc Operational Evaluation Committee, Minutes of the First Meeting, 6 December 1948, NARA RG 27, entry 133, 6: Records of Cloud Physics Ad Hoc Operational Evaluation Committee, 1948–1950.

- in 5 years: 80 percent effectiveness in dispersing supercooled clouds and 50 percent in dispersing supercooled fogs; 10 percent of warm clouds (temperature above freezing) dispersed;
- in 10 years: all supercooled fogs and clouds would be dispersed; 20 percent of warm clouds dispersed;
- in 15 years: no further problems dealing with supercooled clouds; 20 percent of warm clouds dispersed.

The Panel members also recommended that investigations into all phases of cloud physics be continued to increase knowledge of cloud and fog phenomena, including worldwide cloud seeding experiments to include a variety of climate types.[27] These predictions were wildly over-optimistic, perhaps based more on wishful thinking than scientific evidence.

By late 1950, weather control's possibilities had attracted the attention of Air Force General George Kenny, who foresaw global domination by the nation that could produce weather on demand. According to Kenny, "[t]he nation that first learns to plot the paths of air masses accurately and learns to control the time and place of precipitation will dominate the globe."[28] In turn, weather control had caught the attention of members of the US Congress who thought it was about time to start controlling weather control.

On 8 December 1950, Senator Clinton P. Anderson of New Mexico introduced the Weather Control Act of 1951, which would provide for the development and regulation of weather modification and control. The act would create programs that would assist and foster private- and government-conducted research and development, institute government control of experiments and operations to assure the needs of national defense and security, and administer policies and international arrangements to advance weather control. The act would do all of these things, but there was really very little evidence that anyone could control the weather in any way.

According to Anderson, weather was a national defense problem. Weather control could lead to profound changes in people's way of life, with far-reaching benefits to agriculture, industry, commerce, and the general welfare, and therefore it needed to be developed in an orderly manner. Federal oversight was a must. As the nation's population moved West, there was insufficient

---

27 Panel on the Atmosphere—1949 Technological Estimate [Confidential], 14 April 1949, NARA RDB, box 171, folder 2.

28 Quoted in Frederick C. Othman, "Rain of Power," *NY World Telegram*, 12 November 1950, Library of Congress, Manuscript Collections, Papers of Clinton P. Anderson (hereafter LOC Anderson), box 498, folder: Weather Control Inquiries.

water to support their needs. The nation had to increase its water resources, and weather control was the way to get the job done.[29]

However, was a Weather Control Act modeled on the Atomic Energy Act of 1946 the ideal way to address this issue? Scientists, including climatologist Helmut Landsberg, complained that they did not know enough about weather control to attempt to regulate it, and that congressmen should be encouraging research instead. Weather control was not equivalent to atomic energy. When the Atomic Energy Act was passed, scientists knew of the implications of atomic energy and its applications. Writing weather control legislation in 1950, Landsberg wrote, made as much sense as writing atomic energy legislation would have made in 1935.[30]

In hearings that took place on the Weather Control Act and two other weather modification bills introduced at roughly the same time, Anderson remained concerned about the relationship between national defense and weather control. (A large portion of the hearings had actually addressed conflicts between commercial seeders in the Western United States.) According to Anderson, the uncontrolled use of silver iodide could be more harmful to the nation's defense than "five atomic bombs."[31] And while he thought his Weather Control Act was the answer to the problem, the military services did not agree. All three military branches opposed the legislation, albeit for different reasons. The Air Force argued that insufficient progress had been made on weather control to warrant legislation. Since "modest controls" were already in place to prevent unqualified practitioners from scattering silver iodide seeds across the nation, there was no reason to support this bill, which could interfere with Air Force projects.[32] The Navy wanted to see enough control so they could determine seeding's effects, i.e., seeders should be licensed and report on their efforts. Since this bill went much further, they did not support it either. Navy officials did support indemnifying contractors who carried out military research on weather control as long as no willful misconduct was involved.[33] In other words, if a Navy-sponsored experiment resulted in dumping a large

29 Congressional Record—Senate, "Introduction of the Weather Control Act of 1951," 8 December 1950, p. 16475, LOC Anderson, box 498, S. 4236.

30 Helmut Landsberg to Charles F. Brown, 26 December 1950, NARA RDB, box 480, Legislation 1950–51.

31 "Silver-iodide Rain Test called Threat to Defense," *New York Times*, 16 March 1951, p. 25.

32 Robert E.L. Eaton, BGEN, USAF, to Department of the Army, Chief of Legislative Liaison, 15 March 1951, NARA RDB, box 459, folder 8.

33 T.A. Solberg, RADM, USN, Chief of Naval Research to Judge Advocate General, 26 March 1951, NARA RDB, box 459, folder 8.

amount of rain on someone's farm and washing out the crops, the government would pay the court judgment levied against a contractor if he had used the proper procedures and had been sued by the farmer.

Once the hearings had concluded, the military services agreed amongst themselves that they needed restrictions on cloud modification in order to protect their experimental work, but the legislation before Congress was not particularly helpful. What they wanted was to restrict civilian cloud seeding and to continue receiving funding for military research.

When none of the first round of weather control bills passed, a new bill was introduced in October 1951. This bill, sponsored by Senator Francis Case of South Dakota, would establish a temporary committee to examine the results of rainmaking experiments and research. Once its work had concluded, committee members would recommend permanent weather control legislation.[34]

The military services were, once again, unable to present a united front before Congress. The point of contention this time was the timeframe allotted for the advisory committee to provide a full statistical analysis of cloud seeding operations. The services suggested that the committee present an interim report after two years, and a final report two to three years later. In general, they supported the indemnity clause, although the Army thought wider coverage was needed for the contractors conducting the research. The Navy and the Air Force fretted that the bill had no provision for withholding classified information related to rainmaking in the interest of national defense. The Department of Defense and the Army were also concerned that the lack of control over weather modifiers might lead to contamination of their experiments.[35] Once again, the bills did not come up for a vote before the end of the session. Another chance to meet the needs of the military services had been lost.

In January 1953, bills were introduced into both the Senate and the House of Representatives that repeated the bills introduced in the previous ses-

34 Wadsworth Likely, "Bill to Permit Rain-making for 2 years offered," *Albuquerque Tribune*, 10 October 1951, LOC Anderson, box 520, S2225: Rainmaking Bill by Senator Case and others.

35 Frank Pace, Jr., Secretary of the Army, to Rep. Robert Crosser, Chairman, Interstate and Foreign Commerce Committee, 7 August 1951; Frank Pace, Jr., to Hon. Frederick J. Lawton, Director Bureau of the Budget, 8 August 1951, both in NARA RDB, box 459, folder 8; Frank Pace, Jr., to Hon. Edwin C. Johnson, Chairman, Committee on Interstate and Foreign Commerce, 13 March 1952, LOC Anderson, box 520, S2225: Rainmaking Bill by Senator Case and others; Walter G. Whitman memo for Assistant Secretary of Defense for Legal and Legislative Affairs, circa March 1952; Captain E.C. Stephan, USN, Director, Legislative Division, to MGEN Miles Reber, USA, both in NARA RDB, box 459, folder 7.

sion. This time, Senate Bill 285, 83rd Congress, 1st Session (an act "to create a committee to study and evaluate public and private experiments in weather modification") came up for a vote in June 1953, became law on 13 August 1953, and created the Advisory Committee on Weather Control (ACWC).[36] Senator Anderson had not gotten his desired Weather Control Commission, but he had secured an investigatory group that would analyse deliberate weather modification, judge its efficacy, and recommend legislation that would determine the future of weather control as a state tool.

President Dwight D. Eisenhower established the Advisory Committee on Weather Control in late 1953, naming meteorologist and retired Navy Captain Howard T. ('Shorty') Orville chairman. Orville was joined by a number of other officials, as well as representatives from cabinet departments (Defense, Agriculture, Commerce, Interior; Health, Education, and Welfare) and the National Science Foundation. Together they sought out experimental data from weather control research as well as insights from people across the country who had been affected by or had used weather control techniques.[37]

Although input from departments at the cabinet level was generally available to the public, the Defense Department did not openly share its work. An active patron of weather control research and experiments since the mid-1940s, it looked upon weather control as a strategic and tactical tool.[38] Hence, research under its patronage continued in a parallel, classified universe, hidden from not only the public, but usually from other cabinet departments as well. Based on input from the other departments, the ACWC was informed about a wide range of ideas for using weather to mitigate, solve, or prevent a whole host of social and economic problems. Therefore, ACWC members could scarcely have concluded that experimental weather control work should not continue. Press coverage of statements by scientists involved in these efforts that remind one of science fiction novels had already led many people to believe that weather control was just around the corner. During the Cold War, when people often felt at the mercy of powers beyond their control, the thought of people controlling something seemingly as uncontrollable as the weather would have been a strong motivation to continue research into the possible deployment of weather control techniques.

36 S. 285 Union Calendar No. 378, 83rd Congress, 1st Session, 8 June 1953; Public Law 256 83rd Congress, Chapter 425, 1st Session, p. 285, 13 August 1953.

37 "U.S. to do something about the weather," *New York Times*, 10 December 1953, 1.

38 Draft report of the President's ACWC concerning DoD activities in the field, 12 April 1954, NARA Records of the Secretary of Agriculture, Record Group 16, box 2381, folder: Weather Control Advisory Committee, April–August 1954.

Overall, the Advisory Committee on Weather Control concluded that weather control efforts would be best served by funding a multi-year program of experimentation and research. Abandoning their efforts to regulate weather controllers, Senators Case and Anderson, among others, sponsored Senate Bill 86, which would provide for long-range experimental research on cloud modification under the leadership of the National Science Foundation. The program called for a $5 million budget over five years (equivalent to almost $41 million in 2013). This legislation did not, of course, preclude the military services from launching or maintaining their own efforts toward using weather as a weapon. And they proceeded to do so.

Weaponizing Weather in the 1960s

The military services had continued to investigate weather control techniques after Project Cirrus wound down in 1952. By the early 1960s, they were running a number of projects, each with a different goal. The Army's 'Project Flagstaff' was designed to modify clouds so that they could be evaluated physically (by examining changes in their physical structure), and not statistically (by determining if the changes were statistically significant or if they could have happened by chance). The Army was also studying characteristics, causes, predictability, and possible dissipation of whiteout conditions in Arctic areas. It had had some success in dissipating supercooled cloud and fog layers, but needed to find a way to dissipate them if flight conditions were too dangerous.[39] An Army project in Greenland used 'crickets'—small rockets carrying dry ice or chemicals that encouraged fog to form raindrops—to clear fog from runways. This work, coordinated by the Army's Cold Regions Research and Engineering Laboratory in Hanover, New Hampshire, involved using crickets that could lift a parachuted 'payload' of seeding materials to about 1,400 meters and clear a radius of about 200 meters.[40]

The Navy was working on 'Project Blackbody': the use of carbon black to disperse clouds; and more significantly on 'Project Stormfury', which made use of a new pyrotechnic delivery technique, which had been created by the Naval Ordnance Test Station in China Lake, California, to dissipate hurricanes.

39 National Science Foundation, "Weather Modification," Fourth Annual Report for Fiscal Year Ended June 30, 1962, Washington, DC, US Government Printing Office, 1963.

40 Walter Sullivan, "U.S. Tests Rockets in Greenland for Clearing Fog Over Airstrip," *New York Times*, 26 July 1964, p. 19.

Lastly, the Air Force had been concentrating its efforts on warm and supercooled fog and stratus clouds. 'Project Pea Soup' was conducted near Arcata, California, and aimed at dispersing fog. For this purpose, the Air Force dropped 27 kilograms of carbon black per mile from a crop duster on top of stratus clouds that were 183 meters thick. Five of those tests resulted in small clearings. The Air Force was also engaged in identifying an economical way to dissipate supercooled clouds by self-contained seeding systems, which would be carried aboard both transport and cargo aircraft and dissipate clouds with dry ice. In 'Project Blacksheep,' the Air Force continued to work on contrail suppression, which posed a serious problem for the Air Force: in a war zone, jet contrails provide a direct path to the jet for enemy fighters to follow.[41]

The military weather control program that carried the most significant political implications was called 'Popeye.' The classified version of Project Stormfury, Popeye was designed to deploy a classified pyrotechnic device over selected areas of Laos and North Vietnam during the monsoon season, with the intention of keeping the Ho Chi Minh trail muddy and thus hampering the flow of supplies and personnel into South Vietnam. Although Project Popeye had significant military potential, it also had the potential to cause serious political problems. The initial experiments had been very successful, but it was possible that seeding would lead to flooding in populated areas, or wash out crops.[42]

In February 1967, the Joint Chiefs of Staff recommended that the program be extended over the entire length of the Laotian Panhandle and southern North Vietnam. Diplomats at the State Department, who were more attuned to the potential political fallout if civilian casualties resulted from Popeye's cloud seeding, were concerned about using the technique over coastal areas because of possible damage to urban centers or agricultural land. Therefore, they suggested that the technique be limited to critical choke points along the Ho Chi Minh trail to improve the possibility of obtaining data on its military effectiveness and its controllability. Presidential science advisor Donald

---

41 National Science Foundation, "Weather Modification," Fourth Annual Report for Fiscal Year Ended June 30, 1962, loc. cit. Civilian weather control programs, including Skywater (to tap non-precipitating clouds floating over the United States) and Skyfire (to snuff out forest fires and limit lightning), were underway at the same time. See Kristine C. Harper, "Climate Control: United States Weather Modification in the Cold War and Beyond," *Endeavour* 32 (2008), 20–6.

42 For details on the Johnson administration's use of weather control in India and Vietnam, see Ronald E. Doel and Kristine C. Harper, "Prometheus Unleashed: Science as a Diplomatic Weapon in the Lyndon B. Johnson Administration," *Osiris* 21 (2006), 66–85.

Hornig explained to President Lyndon Baines Johnson that the political problem could explode if people found out about Popeye, and the US were charged "at home and abroad with initiating a new kind of warfare—'weather warfare'." Although Hornig viewed using heavy rains to wash out jungle trails as a humane weapon, it would mean that the US was the first to open a 'Pandora's box' of unknown threats to future civilizations.[43]

The press and members of the scientific community had already been discussing the possibilities of weather warfare: melting polar icecaps, modifying world weather conditions to the disadvantage of large areas, or creating or steering hurricanes that would impact enemies. Weather control techniques could be used to flood, and hence destroy, agriculturally intensive areas. As Hornig reminded Johnson, an earlier report issued by the National Academy of Sciences and scientific testimony before Congress had urged that weather control be used for peaceful purposes only. If international reaction were adverse, nations could decide to withhold weather data. At that time, only North Vietnam and Albania, which were geographically very small, were not reporting to the world-wide system of weather data collection centers. But numerical weather prediction models needed all available data in order to predict the weather. Whatever the White House decided, it was important to keep the project secret. As Hornig reminded Johnson, recent attempts to use the same techniques on a trial basis in India in an effort to break the serious drought conditions in Bihar and Uttar Pradesh provinces could expose the project to outside scrutiny that they did not need.[44]

The Vietnam and Laos weather control project (now known as 'Project Compatriot') continued into mid-May of 1967. Military personnel had expended 234 seeding units, a smaller number than in previous weeks due to a lack of seedable clouds. They reported the presence of standing water, mudslides, and muddy, rutted roads. Major rivers were running so high that they could carry river traffic for their entire length.[45] Johnson's national security advisor, Walt Rostow, filled the president in via a 'Top Secret—Literally Eyes Only' memorandum on 22 May 1967, in which he forwarded a military plan to induce rainfall that would flood out the Song Ca basin, thus impeding the movement of supplies to South Vietnam in an area south and west of Vinh. However, these same

43 Hornig to LBJ, 20 February 1967 [Top Secret], Lyndon Baines Johnson Presidential Library and Archive, Austin, TX, National Security Files (hereafter LBJ NSF), Country File Vietnam 41/Vietnam Memos (B), Vol. 66, 2/17–28/67.

44 Ibid.

45 Project Compatriot Weekly Report #9 [Top Secret "Eyes Only"] for 12–18 May 1967, 18 May 1967, LBJ NSF, Country File Vietnam 88/Vietnam 3P Project Compatriot 5/67–7/67.

efforts were underway in a secret US government-initiated program in India, and they needed to be able to mask the abnormal rainfall. The target area in Vietnam experienced extremely variable rainfall from year to year, so these heavy rains were unlikely to arouse suspicion. However, the US government needed to prepare an explanation in case the North Vietnamese accused it of being responsible. The lack of definitive proof meant that such accusations would not be taken seriously "by responsible persons." Nevertheless, Rostow told Johnson, "the validity of such a possible accusation, even if unsupported by evidence at the time, taken with the probability that the fact that we had the capability will be revealed at a later date, require that our response be carefully considered."[46] Rostow continued:

> It should be noted that if we carry out and defend the Song Ca operation on these lines, we would be considered by many to be drawing the line applicable to future situations—that we did *not* consider weather warfare justified if it did involve significant damage to civilian life and health. We need to think for a moment whether we really mean this in relation to a major conflict. Would we, for example, have rejected partial flooding of the Netherlands if they could have got us to Berlin two months sooner? Whatever the form of these future questions, the short answer may be that we can face them when they arise, and that at that point the international lawyers and precedent-seekers will be less important than other factors such as the nature of the opponent.[47]

The following day, Hornig told Rostow that they should know more about the expected military effect of cloud seeding in quantitative terms. There were several possible adverse factors: continuous flooding would be a hazard to life, health, and sanitation, which could create food shortages for the very young, aged, and infirm. Hornig considered the risk of a security leak from informed parties low, but it was possible that the unusual cloud formation signatures on Saigon weather radars might betray the program. When questions had arisen about the cloud formations, administration spokesmen had dismissed them as "early monsoon conditions." However, since unclassified weather satellites could see them, meteorologists were certain to notice and comment on them in the future. Hornig argued that, "[t]he degree of revulsion to be expected in the domestic and international meteorological circles at the initiation of

46 Rostow to LBJ [Top Secret—Literally Eyes Only], 22 May 1967, LBJ NSF, Country File Vietnam 88/Vietnam 3P Project Compatriot 5/67–7/67.

47 Ibid.

'weather warfare' should not be underestimated, especially since our highly disaffected scientific community, at least in the universities, is ready to join in the chorus." He continued, "[n]or should we underestimate the potential damage to our world position of our being the first to initiate a new form of warfare. Not only because it is in the sequence—atomic bomb, riot gases, defoliation, napalm—but because of the picture it may give of a nation flailing out with every tool at is disposal—particularly if it should prove ineffective."[48]

Despite the concerns expressed by both Hornig and Rostow, the use of weather control techniques continued in North and South Vietnam, and Laos, and were eventually extended into Cambodia. They only stopped two days after journalist Seymour Hersh broke the story in the *New York Times* in mid-1972.[49] During the congressional hearings that followed, Defense Secretary Melvin Laird denied that the United States was using weather as a weapon. With the change of administration—Richard M. Nixon had been sworn in as president in January 1969—no one had informed Laird about the weather weapon or the fact that Johnson had approved its use. Laird subsequently apologized to members of Congress for his mistaken testimony. But in the meantime, the US Senate passed a resolution urging the ban of environmental weapons, and in summer 1975, the United States and the Soviet Union agreed not to use weather as a weapon.[50]

Conclusion

The atmospheric sciences benefitted greatly from Cold War funding. Meteorologists were aided by numerical weather prediction, climate modeling, weather satellites (both geostationary and polar orbiters) with a variety of remote sensors, improved radar, more sophisticated computers for better model output, automatic weather station instruments of all kinds that provided data from remote and unmanned sites, and improved weather facsimile machines. All of these increased their ability to both understand atmospheric processes and provide improved weather services to military and civilian customers. However, all of the additions to meteorology's scientific arsenal likely

---

48 Hornig to Rostow, 23 May 1967 [Top Secret] in response to Rostow to Hornig, 23 May 1967, LBJ NSF, Country File Vietnam 88/Vietnam 3P Project Compatriot 5/67–7/67.

49 Seymour M. Hersh, "Rainmaking is used as Weapon by U.S.," *New York Times*, 3 July 1972, p. 1.

50 Bernard Gwertzman, "A U.S.-Soviet Ban on Weather Use for War is Near," *New York Times*, 24 June 1975, p. 1.

would have occurred in the absence of the Cold War as well. They just might have taken longer to appear because funding would not have been as readily available to the civilian-run Weather Bureau as to the military's meteorological services. In this sense, these meteorological advances resulted from scientific undertakings during the Cold War.

However, without the Cold War and concerns over unknown Soviet capabilities to control the weather, initial military-funded research focused on weather control would have been unlikely. Langmuir would not have received five years of support for Project Cirrus, and the military services would have been more concerned about producing forecasts than trying to modify the weather. Had Langmuir continued with his research without military support, it is possible that commercial-sector consulting meteorologists may have adopted cloud seeding techniques and sold them to farmers, ranchers, utility companies, ski resorts, and other users, just as they did during the Cold War. However, this does not seem very likely, especially not in the 1950s. Most meteorologists had expressed serious reservations about the process, so that both commercial and military applications of weather control to specific meteorological problems might have taken place on a smaller scale only. Appeals to needs for additional water in the Western US would have been met by traditional techniques: the building of dams and reservoirs on rivers and streams, rather than expensive attempts to modify the weather across the entire continent. Therefore, while meteorology itself was a science advanced by the Cold War, the weather control portion that often operated outside of the meteorology community is better seen as Cold War science that was completely dependent upon the Cold War mindset of using scientific and technological fixes to maintain geopolitical dominance for its initial survival and continued development.

During the 1950s and 1960s, when weather control efforts by the US government were at their peak, the majority of meteorologists affiliated with the Weather Bureau and academic departments in universities pointed out that the results seen by Langmuir and his researchers at General Electric fell within the range of typical weather.[51] Additional experiments, including those conducted by the Artificial Cloud Nucleation Project in the early 1950s, and research reviews such as those conducted by the Advisory Committee on Weather Control (ACWC) and the National Academy of Sciences' Panel

51 USWB Scientific Services Division, "Evaluation of Dr. Langmuir's Claims of Producing Weekly Cycles in Weather by Cloud Seeding," 6 June 1952, NOAA Central Library, Silver Spring, Maryland.

on Weather and Climate Modification, all produced inconclusive results.[52] At times their seeding produced positive, measurable results, and at other times it did not. Statistical tests did not usually show significant results.[53] These were tough to do in any case, as it was hard to establish whether a randomly selected cloud produced precipitation due to seeding or not. Into the late 1970s, scientists were still finding that "the experimental evidence for cloud seeding has not yet reached the levels of objectivity, repeatability, and predictability required to establish new knowledge and techniques."[54] In 2003, the US National Research Council looked into weather modification and control once more, and concluded that it had "little reason to differ" from earlier findings.[55] While quality atmospheric science has been conducted since the mid-1940s as a result of attempts to control the weather (particularly work in cloud physics and dynamics, and in advanced modeling techniques) the ability to control weather over large expanses of Earth's surface envisioned in the mid-twentieth century has not followed. The interactions of the physical, chemical, and dynamical attributes of the atmosphere are extraordinarily complex and scientists are still unlocking their secrets with the aid of advanced remote sensing equipment and massive computers. Despite the on-going discussions of possible geoengineering solutions to Earth's current and future weather and climate problems, focusing on the science and less on possible quick-fix applications appears to be the better choice.

---

52 See, for example, Advisory Committee on Weather Control, "Final Report of the Advisory Committee on Weather Control," Washington D.C., 1957; Sverre Petterssen, "Cloud and weather modification," *Meteorological Monographs* 2 (1957), 111; National Research Council, *Scientific Problems of Weather Modification* (Washington D.C., 1964).

53 T.A. Jeeves, L.M. LeCam, J. Neyman, and E.L. Scott, "Problem of Documentary Evidence of the Effectiveness of Cloud Seeding," unpublished report from 1952, used as evidence in a report to Howard T. Orville on 10 April 1954, Library of Congress, Papers of Irving Langmuir, box 11, folder 3; Carleton E. Buell, "An Approach to the Evaluation of the Results of Rainmaking: A Progress Report," presented at the Symposium on Cloud Seeding, Institute of Mathematical Statistics, East Lansing, Michigan, 4 September 1952.

54 Weather Modification Advisory Board, "The management of weather resources: Proposals for a national policy and program. Final report of Weather Modification Advisory Board," (Washington, D.C., 1987), vol. 1, 35.

55 Committee on the Status and Future Directions in US Weather Modification Research and Operations, National Research Council, "Critical Issues in Weather Modification Research," (Washington D.C., 2003), 21. For more information on geo-engineering, see James Rodger Fleming, *Fixing the Sky: The Checkered History of Weather and Climate Control* (New York, 2012).

CHAPTER 11

Small State versus Superpower: Science and Geopolitics in Greenland in the Early Cold War

*Matthias Heymann, Henry Nielsen, Kristian Hvidtfelt Nielsen and Henrik Knudsen**

This article will investigate the relationship between the small state Denmark and the superpower USA by looking at the role of science and technology in Greenland in the early Cold War. Many features of this relationship remained hidden until a catastrophe hit in the north of Greenland. On 21 January 1968 a B-52 bomber of the US forces crashed into the sea ice close to Thule. The bomber carried four thermonuclear bombs, which disintegrated upon impact. Plutonium was dispersed onto a large area. However, the thrust of this crash was at its most powerful in Washington and Copenhagen: it put an end to an era of complicated and fragile diplomatic arrangements. As soon as the information of the Thule accident reached the Strategic Air Command and the Pentagon, contingency plans were executed all over the United States. The US Air Force "immediately sent experts to recover the weapons and to survey the damage." It also sent experts on radiation and the health impacts of radiation.[1] Within a few hours, an 'air bridge' was established from the USA to Greenland to ship experts and supplies to Thule. Soon after, the Danish government sent high-ranking scientists and specialists to assist. The US Department of Defense launched a large and continuously expanding clean-up operation, Project Crested Ice, which operated until September 1968. Hundreds of experts and staff, and several tons of cargo were flown in to Thule.

At the height of its existence, Project Crested Ice involved more than 500 staff members, and was divided into stages. The first stage of the operation was dedicated to the investigation and detection of radioactive contamination, and the recovery of aircraft debris and weapons parts. Investigations of the crash site on the ice revealed significant contamination of the debris, snow and ice,

* We are grateful to the Danish Carlsberg Foundation for the funding of our research. Matthias Heymann would also like to thank the Rachel Carson Center, University of Munich, for supporting his writing of this article with a fellowship.

1 Strategic Air Command (ed.), *Project Crested Ice: The Thule nuclear accident*, vol. I, SAC Historical Study No. 113 (1969), p. 3.

 | DOI 10.1163/9789004264229_012

which also affected the rescue teams and equipment. Decontamination of staff and equipment was a daily exercise. Radiation scientists immediately started their surveys; they located and marked a so-called 'zero-line' around the impact area, which indicated the line of zero contamination, and confined a contaminated area of about three by one miles.[2]

In mid-February a second part of the operation began: the decontamination of the accident area and further scientific investigations of the surrounding environment. To accomplish this task, an expanded civil engineering division was established on 1 March.[3] Burning fuel had created a large blackened area on the ice, caused by melting snow that had formed a crust after re-freezing. This crust "contained a tremendous amount of contamination,"[4] and had to be removed, together with the contaminated snow surface within the entire contamination area. US Air Force personnel used graders to collect the contaminated snow and ice at the crash site, while US and Danish servicemen would then load the same into wooden boxes. These were moved to a holding area near Thule Air Base known as the 'Tank Farm,' and filled into steel tanks which were to be removed from Greenland later by ship.

Project Crested Ice involved a range of scientific investigations to analyse the distribution and impact of radioactive materials, including ice core analyses, hydrographic surveys (for the purpose of which a mobile oceanographic laboratory was shipped to Thule), and investigations of the bottom of the sea with a small submarine.[5] The entire effort is estimated to have cost some $9.4 million, with a total over seventy military and other US agencies involved.[6]

---

2 Robert E. McElwee, *Project Crested Ice, USAF B-52 accident at Thule, Greenland, 21 January 1968*, Report by the Defense Atomic Support Agency Nuclear Emergency Team (1968), pp. 2–3.

3 G.S. Dresser, "Host base support," *USAF Nuclear Safety* 65 (1970), 29.

4 Joint Committee on Atomic Energy, Congress of the United States, Minutes of the Executive Meeting No. 90-2-11, Wednesday, 20 March 1968, p. 15, accessible online at: http://upload.wikimedia.org/wikipedia/commons/e/e4/Documents_Concerning_Thule_Accident.pdf [accessed 4 April 2014] (hereafter Joint Committee, Minutes).

5 McElwee, *Project Crested Ice*; Børge Fristrup, "Ice investigations," *USAF Nuclear Safety* 65 (1970), 84–6; Marshall E. Neal and Weston A. Roe, "North Star Bay Oceanography," *USAF Nuclear Safety* 65 (1970), 91–5.

6 Stephen I. Schwartz, *Atomic Audit, The Costs and Consequences of U.S. Nuclear Weapons Since 1940* (Washington DC, 1998), p. 410.

FIGURE 11.1 *The end of 'Project Crested Ice:' Colonel C.S. Dresser, US base commander at Thule Air Base (right) and Commander Jørgen Mølgaard, Danish Liason Officer, demonstrate the completion of the collaborative cleaning effort on Friday, 13 September 1968 (C.S. Dresser, "Host base support," in:* USAF Nuclear Safety *65 (1970), p. 31).*

Scientists played a crucial role in Project Crested Ice: not only as experts for the cleaning process, but also as diplomatic agents to mediate between both countries involved, disperse the political irritation this event had caused, and to contain the embarrassment of the scandal this crash represented on both sides of the Atlantic. The US Air Force worked intensively with Danish scientists to plan the clean-up operation. According to US scientists "no biological hazard existed as a result of the crash"[7]—according to them, the plutonium could be left where it was without causing much of a hazard.[8] The Danish experts, however, disagreed and did not accept the abandonment of contaminated materials in the bay: all decently sized pieces of debris, as well as snow and ice, must be picked up and hauled away. At a meeting of the groups of US and Danish scientists on 15 and 16 February in Copenhagen both reached a pragmatic understanding about the extent of the planned clean-up operation. Both sides agreed that the project should not cause "a hazard to people or biological species," but "an effort will be made to remove the main part of the radioactivity which is on the ice."[9]

US officials commented repeatedly on the friendly and cooperative behavior of the Danish scientists. The Danish scientists shared a 'gentlemen's understanding' with the US officials about the issues at hand and agreed to a pragmatic compromise. The Historical Study of Project Crested Ice, which was commissioned by the US Strategic Air Command, valued the cooperative attitude of scientists on both sides even higher than their scientific achievements. "Perhaps their most important contribution [...] was their day-to-day working association with the Danish scientists who came to Thule. The Danes, a highly intelligent and skilled group, held all the cards, yet they chose not to be difficult or demanding."[10] Also, for Denmark, much was at stake. In the Congress Hearings, US Secretary of Defense, Carl Walske explained that the Danish scientists "realized they had a public relations problem and a good job should be done in cleaning up what could be cleaned up."[11] In fact, Danish scientists had to maneuver difficult waters and were forced to consider many factors, including the advantages of a comprehensive clean-up to avoid environmental consequences, the security and consent of the local Greenlandic population, rising diplomatic tensions with the US, and appeasing the Danish public while avoiding to draw attention to US activities in Greenland.

---

7 Strategic Air Command, *Project Crested Ice*, p. 27.

8 Joint Committee, Minutes, p. 26.

9 Joint Committee, Minutes, p. 35.

10 Strategic Air Command, *Project Crested Ice*, pp. 25, 64.

11 Joint Committee, Minutes, p. 36.

The Thule crash was as much a political scandal in Denmark as it was a military scandal in the USA. It unveiled top secret military operations in the Arctic, which was a disaster for the US Air Force and the US government and, at the same time, a huge embarrassment to the Danish government just four days before national elections. The crashed B-52 bomber was part of Operation Crome Dome, a highly secret Airborne Alert program consisting of continuously airborne US bombers flying, on alert and around the clock, while carrying nuclear bombs. One part of the Airborne Alert program was the Thule Monitoring Mission, the continuous airborne surveillance of Thule Air Base, of which the crashed B-52 formed a part. This assignment was top secret and pursued without the knowledge of civilian authorities in the United States or Denmark; only a very small number of officials at top level in the Danish government, including Prime Minister Jens Otto Krag, had been informed.[12]

In 1957, Denmark had decided on a ban of nuclear weapons on Danish territory, waters and airspace. In practice, however, Danish governments had operated with two nuclear policies: an open and official policy that covered continental Denmark and enacted the nuclear ban on one hand, and a tacit and unofficial policy that covered Greenland while not enforcing the nuclear ban—certainly unknown to parliament and the public—on the other. This bilateral policy originated in 1957 with a letter from Danish Prime Minister Hans Christian Hansen to the US Ambassador in Copenhagen; it was silently continued until 1968. In November 1957, the US Ambassador wrote to Hansen, asking whether the Danish government demanded to be informed about US plans to store nuclear weapons in Greenland. Hansen replied in a private and highly confidential letter a few days later, stating that these remarks "do not [...] give cause to any comments from my side."[13] The Danish Prime Minister thus decided to turn a blind eye to US nuclear activities in (and above) Greenland, and to avert his gaze for the years to come. This policy was more than a red herring in Danish domestic politics. It was a silent consensus with the US government, which relied on and repeatedly reminded the Danish government about the same.[14]

---

12 Scott Douglas Sagan, *The Limits of Safety: Organizations, Accidents, and Nuclear Weapons* (Princeton, NJ, 1993), pp. 167–70; Thorsten Borring Olesen, "Tango for Thule: The Dilemmas and Limits to 'The Neither Confirm nor Deny Doctrine' in the Danish-American relationship, 1957–1968," *Journal of Cold War Studies* 13 (2011), 116–47.

13 Nikolaj Petersen, "The H.C. Hansen Paper and Nuclear Weapons in Greenland," *Scandinavian Journal of History* 23 (1998), 21–44, on p. 22.

14 Olesen, "Tango;" Thorsten Borring Olesen and Poul Villaume, *I blokopdelingens tegn, 1945–1972. Dansk udenrigspolitiks historie, vol 5.* (Copenhagen, 2005), pp. 635–40.

When HOBO 28 exposed the Danish deceit, the Danish government was put into a dilemma: it risked to be accused of having betrayed the Danish public about Denmark's nuclear policy for more than a decade, but also to cast light on the extent of US activity in Greenland of the past two decades. This had been kept as secret as possible from the Danish public. A revelation of the fact that Denmark had largely ceded military sovereignty over Greenland to the USA would certainly have an unpleasant impact. Prime Minister Jens Otto Krag resorted to an artful lie as soon as the day after the Thule crash. In the late afternoon of 22 January, keeping the public in the dark for the day, he published a press release which stated explicitly that "in agreement with the government policies, there are no nuclear weapons within Danish territory. This also applies to Greenland; and therefore overflight of Greenland by aircraft with nuclear bombs cannot take place."[15] This press release contained two messages at the same time: it was, firstly, intended to calm the concerns of the Danish public and media; and secondly, it contained a hidden, bold message to the US government: Krag was revoking the silent agreement with the US government to accept nuclear weapons in or over Greenland. He succeeded in containing the internal Danish scandal by trading it in for a full-blown diplomatic crisis. The US government immediately understood the hidden message and responded with strong irritation. US Secretary of Defense Robert McNamara decided on the same day to cancel all Thule Monitor Mission flights involving nuclear weapons.[16]

The Thule accident forced US military engagements in Greenland and Greenland's strategic role in the Cold War into the public eye. It also revealed the delicate balance between dependence and cooperation that formed the relationship between the superpower USA and the small state Denmark. This cooperation was not always friendly. It gave rise to anxiety on both sides of the Atlantic. The crash also cast light on the role of Greenland science for Denmark and the USA. Science and technology proved not only crucial for the management of an extremely hostile environment in Greenland, and for coping with challenges such as the Thule accident. Science and technology also served as instruments for the pursuit of diplomatic purposes, strategic needs, and geopolitical interests. For both states, Denmark and the USA, science was a vital means for gaining and retaining several types of control: the physical control of the arctic environment, which would be a likely warzone in case of an

15 Olesen, "Tango," p. 132.

16 Sagan, *Limits*, p. 193; Olesen, Villaume, *I blokopdelingens tegn*, p. 641. McNamara had not been in favor of the Airborne Alert program since 1966. The program was fully cancelled by July 1968; Sagan, *Limits*, pp. 178–9, 195.

actual conflict between the USA and the USSR, as well as political and military control of the entire island. This article will show that the size of a country has a significant bearing on perceptions and policies in world affairs.[17]

How did the superpower USA pursue its military-scientific interests and manage the idiosyncrasies of the small state Denmark? And how did the small state Denmark defend its interests and respond to challenges caused by the superpower USA? We will discuss three cases of scientific projects in Greenland that illustrate the workings of the unbalanced power relations and changing responses to the same. First we look at the operation of Greenland's weather stations in the immediate post-war years; then, at the establishment of 'Camp Century,' the city under the ice, in the late 1950s; and finally, we discuss 'Operation PCA 68', the ambitious (but futile) American rocket project of 1968.

Weather Stations: New Strategic Imperatives vs. Sovereignty Concerns

On 13 June 1946, Danish meteorologist Viggo Laursen, officer in charge at the meteorological observatory in Ivigtut, South-West Greenland, received a letter from Copenhagen. Director Helge Petersen, head of the Meteorological Institute, informed him that he was to be relocated to Thule, about 2000 kilometer further North. Laursen would be responsible for building up a new magnetic observatory.[18] In a meeting on 20 June in Ivigtut, Director K. Oldendow, head of the Greenland Administration in Copenhagen, explained the rationale behind this mission to Laursen: the Americans, so Oldendow, intended to establish a new and larger meteorological observatory in Thule than the present one, equipped for the investigation of upper air observations and of the polar magnetic field. But the Danish Ministry of Foreign Affairs and the Greenland Administration "ascribed for political reasons [...] great importance [to the aim] that it was Danes who established and operated this station." The USA had already begun to send staff to Thule. "Denmark had no opportunity to prevent this," Oldendow explained, "but the idea now was

17 The general argument has been made in Matthias Heymann and Janet Martin-Nielsen, "Introduction: Perspectives on Cold War Science in Small European States," *Centaurus* 55 (2013), pp. 221–42.

18 Helge Petersen to Viggo Laursen, 13 June 1946, Danish National Archive, Copenhagen, 1956 Meteorologisk Institut, box 1: 1945–1952 USA's forhold til danske vejrstationer (Thule), folder 2 Thulesagen 1945–46 (hereafter DNA, Thulesagen).

that we should try to neutralize the American effort by sending our people to Thule." Oldendow did not forget to add that any information he provided was "top secret."[19] On 15 July, Petersen sent another letter to Laursen. He reported that the negotiations were difficult, and his impression was that discussions of magnetic investigations were rather put on the back burner. "And even if it is of course of great interest to get a permanent observatory up in Thule, it is, I believe, *not* the right time for it. However, a lot of pressure was put on me, so that it seemed to me that we were compelled to give in, even if it is unpleasant both for you and for the institute."[20]

Laursen was not at all happy with the order he had received. He would have preferred to continue his work in Ivigtut. The build-up of a new station in Thule was not a feasible plan from a scientific point of view, but rather, a political move. From one day to the next, Laursen was entrusted with a diplomatic mission (as he learned from Oldendow, who asked him to represent Danish interests), and expected to act as liaison officer to the American forces that were still present in Greenland. For this purpose, Oldendow had instructed him to familiarize himself with all important speeches and political writings on Greenland policies. In addition, Laursen was asked to deal with an issue which the Danish Government regarded a very sensitive one: the plight of the indigenous Greenlandic population. "It was first and foremost of utmost importance that Thule's [indigenous] population would be saved from the deleterious influence, which inevitably was the consequence of such numerous American and Danish people's presence at this place," Oldendow explained.[21]

Laursen's story is only a very small part of extended and complicated negotiations about weather stations in Greenland.[22] After Denmark's German occupation in World War II, Denmark was not in a position to administrate Greenland. Instead, the USA took over the defense and supply of the island.

---

19 Referat af Samtale med Magister Laursen i Ivigtut, 20 June 1946, DNA, Thulesagen. All Danish quotations have been translated by this article's authors.

20 Helge Petersen to Viggo Laursen, 15 July 1946, DNA, Thulesagen.

21 Referat af Samtale med Magister Laursen i Ivigtut, 20 June 1946; Helge Petersen to Viggo Laursen, 13 June 1946, DNA, Thulesagen.

22 For a full account of this story, see Matthias Heymann, "In search of control: Weather stations and weather data in the early Cold War," (manuscript, Aarhus, 2014), forthcoming in: Matthias Heymann, Kristine C. Harper (eds.), *Exploring Greenland: Expanding Technologies and Creating Geophysical Knowledge in the Cold War* (New York: Palgrave Macmillan). See further: Dansk Udenrigspolitisk Institut (DUPI), *Grønland under den kolde krig. Dansk og amerikansk sikkerhedspolitik 1945–68* (Copenhagen, 1997), pp. 66–87; Niels Amstrup, "Grønland i det amerikansk-danske forhold 1945–1948," in Niels Amstrup and Ib Faurby (eds.), *Studier i dansk udenrigspolitik* (Aarhus, 1978), pp. 155–98.

Based on the 'Agreement relating to the Defense of Greenland'—signed by the Danish Ambassador in Washington, Henrik Kauffmann, in April 1941[23]—the US Forces built bases in Greenland, used the island as a base for the transfer of aircrafts to the war in Europe and, as part of its larger assignment, built and operated weather stations. After the war, Denmark immediately attempted to regain control of Greenland. The Danish government expected US forces to leave the island as quickly as possible and attempted to accelerate this process. The USA, however, had no intention to follow Denmark's wishes, but developed a strong interest in expanding its engagement—even though it was on foreign ground. The 1941 agreement stated in its last paragraph that it remained valid "until agreement has been reached that current threats to the peace and security of the American continent have ended"—a formulation that the Danish government clearly interpreted differently than the US government. In December 1946, the Danish Minister of Foreign Affairs, Gustav Rasmussen, visited Washington to settle the Greenland issue with Secretary of State James Byrnes. Rasmussen was taken by surprise when Byrnes confronted him with the American plans. Byrnes presented three alternatives, emphasizing that the USA were strongly in favor of the final option: (1) A 99 year base agreement between the USA and Denmark; (2) free access for the USA to take over the defense of Greenland on a permanent basis; (3) the acquisition of Greenland for 100 million US dollars.[24] Rasmussen, shocked, immediately declined all three offers, left Washington without reaching an agreement with the Americans, and even without an idea of what a possible compromise, if any, would entail. Years in a state of tension and legal insecurity ensued.

New strategic imperatives of the Cold War moved the Arctic, and in particular Greenland, into a central geopolitical position. Located right between the superpowers, Greenland became a hot spot of strategic interest.[25] The Strategic Air Command (SAC) in particular, which gained a predominant role in the development of the American nuclear strategy, was an underlying factor of the American interest in Greenland.

While the US government and the US Forces prepared highly secret reports about the future role of Greenland, the issue also featured prominently in the

---

23 The agreement is reprinted in DUPI, *Grønland under den kolde krig*, vol. 2, pp. 7–22. Kauffmann acted essentially on his own, independent from occupied Copenhagen.

24 Bo Lidegaard, *I Kongens navn: Henrik Kauffmann i dansk diplomati 1919–58* (Copenhagen, 2005), pp. 410–2.

25 DUPI, *Grønland under den kolde krig*; Nikolaj Petersen, "SAC at Thule, Greenland and the U.S. polar strategy," *Journal of Cold War Studies* 13 (2011), 90–115. For the case of Norway: Rolf Tamnes, *The United States and the Cold War in the High North* (Dartmouth, 1991).

press. On 27 January 1947, Time Magazine published a report on "Deepfreeze Defense," i.e., the role of the Arctic for defense of the continental USA. "Greenland's 800,000 square miles make it the world's largest island and stationary aircraft carrier," it emphasized. "It would be as valuable as Alaska during the next few years, before bombers with a 10,000-mile range are in general use. It would be invaluable, in either conventional or push-button war, as an advance radar outpost. It would be a forward position for future rocket-launching sites. In peace or war it is the weather factory for northwest Europe, whose storms must be recorded as near the source as possible."[26]

The strategic interest of the USA in Greenland put Denmark under strong pressure. The Danish government's policy was guided by a number of serious concerns: it aimed to maintain and defend Danish sovereignty, keep control over the socio-economic and cultural development of Greenland, limit and contain US presence in Greenland (if its presence was, indeed, unavoidable), and control the American influence on Greenlandic society while keeping information about military activities in Greenland, produced by the USA, secret from the Danish public.[27] A set of policies was drawn up for these purposes. Denmark resumed and strongly increased its efforts and presence in Greenland. The modernization of the country and the scientific study of its unique environment were some of the Danes' most important strategies.[28]

The main and most contentious issue in US-Danish relations in the immediate post-war years was the future of the weather stations in Greenland, which the USA had operated during the war.[29] Immediately after the war, the Army Weather Service in Greenland was anxious to turn over weather stations to Denmark, because of "the growing shortage of Army personnel."[30] In July 1945,

---

26 "National Affairs: Deepfreeze Defense," *Time Magazine*, 27 January 1947, accessible online at: http://content.time.com/time/magazine/article/0,9171,778870,00.html [accessed 24 August 2013].

27 For a detailed analysis of the sovereignty issue in the Danish-Greenlandic relationship in the period of 1945–1954, see Erik Beukel, Frede P. Jensen and Jens Elo Rytter, *Phasing out the Colonial Status of Greenland, 1945–54: A Historical Study* (Copenhagen, 2010).

28 Kristian Hvidtfelt Nielsen, "Transforming Greenland: Imperial Formation in the Cold War," *New Global Studies* 7 (2013), pp. 129–54; Nikolaj Petersen, "The Politics of US Military Research in Greenland in the Early Cold War," *Centaurus* 55 (2013), pp. 294–318.

29 The following paragraphs are based on Heymann, "In search of control."

30 Dispatch No. 129, Legation of the United States of America: Josiah Marvel, "Negotiations concerning Danish Manning of Greenland Weather Stations," p. 1, 18 June 1946, National Archive and Records Administration II, College Park, Maryland, Record Group 84, Records of the Foreign Service Posts of the Department of State, Denmark, US Embassy Copenhagen (hereafter NARA, US Embassy Copenhagen), box 9; minutes [December

Denmark was asked to take over five of 24 weather stations, and to send 24 people to staff them by August 1. US forces offered to train the Danes and pay their salaries. Denmark, however, had tremendous difficulties in finding appropriately qualified staff. Twelve operators were made available by the Danish Navy for Greenland in the middle of August. The Danish District Governor in Greenland, Carl Frederic Simony, telegraphed Copenhagen in early October to state that "the Americans were very disappointed."[31] On 17–21 October, Simony and representatives of the US forces met in Godthaab, Greenland, to negotiate a contract about the hand-over of weather stations to Denmark. The contract, signed on 22 October, stated that Denmark would appropriate the first five weather stations by 1 July 1946, and detailed the exact number of Danish personnel to be made available, as well as the number of daily observations to be recorded, and the salaries of Danish personnel to be paid by the US forces.[32] In March 1946, however, the Americans suggested an accelerated process: 18 weather stations were to be operated by the Danish, while the USA would man the remaining six stations. When it became clear at the end of April 1946 that Denmark's preparations lagged behind, the responsible US officers were distressed. Brigadier General C.V. Haynes, Commanding Officer in Newfoundland, who had led the negotiations with Simony in Godthaab, decided immediately—and without the consent of the Danish government—to continue negotiations in Copenhagen to come to a binding agreement with the Danish government.[33]

Difficult negotiations ensued. The Danish government was not well prepared and had not yet decided on its position.[34] The USA demanded to keep all 24 weather stations in operation. This suggestion conflicted with the plan of the Danish Meteorological Institute to reduce their number to only 14 stations: a plan that had been presented to the conference of the Provisional International Civil Aviation Organization (PICAO), in preparation for post-war civil aviation, in Dublin in March 1946. It had been deemed sufficient for the requirements of civil aviation and accepted by the delegates. While the

1945], pp. 3–4, NARA, US Embassy Copenhagen, box 3, Classified General Records 1945–1961.

31 Minutes [December 1945], p. 7, NARA, US Embassy Copenhagen, box 3.

32 Dispatch No. 26, 24 October 1945 by Arvid Holm, American Vice Consul, Godthaab; Contract between the United States of America and the Greenland Administration, p. 2, 22 October 1945, NARA, US Embassy Copenhagen, box 3.

33 Dispatch No. 129, loc. cit., p. 3.

34 For the internal Danish negotiations by the Ministry of Foreign Affairs, the Greenland Administration, the Ministry of the Navy, and the Danish Meteorological Institute, see Amstrup, "Grønland," pp. 170–6.

Meteorological Institute preferred to adhere to this plan, the Danish Ministry of the Navy expressed doubt whether Denmark would be able to find personnel for the operation even of 14 weather stations. The Danish Ministry of Foreign Affairs, by contrast, held a very different view. In a meeting with the US representatives on 27 May 1946, the Ministry emphasized that Denmark intended to take over all 24 weather stations. The legal counsel at the Danish Ministry of Foreign Affairs, Max Sørensen, had recommended a full takeover of all US installations in order to gain full sovereignty and control—this would also provide ample reason to send the US forces home.[35] The ministry even refused the financial support offered by the USA. A note sent to the American Legation[36] in Copenhagen two days later clarified this position, stating "that the Danish Government had already taken steps to provide the personnel for 18 stations and indicated that, contrary to the [US] Army's original idea, the Danish Government would be responsible for paying and supplying its people." The Danish government was not only keen on "participation in the weather reporting system" but sought to gain "independent control over such reporting [...] with a view to assuming, as soon as possible, full responsibility for the entire Greenland weather network." The US Legation was frustrated to see this note. A dispatch to the State Department concluded that it "left a good deal to be desired in terms of a workable agreement about which there would be a minimum of misunderstanding."[37]

The situation had become even more complicated due to American plans to expand weather observation. In April 1946, the US Weather Bureau requested permission for the establishment of a new large weather station at Thule, in the northwest of Greenland (which soon afterwards caused the Danish Meteorological Institute to send Viggo Laursen to Thule). The Danish government responded to this request as it had done before. In a discussion at the Ministry of Foreign Affairs on 25 May the American Legation merely received the information that "the Danish Government was ready to consider further recommendations in the interest of general weather reporting and, for this purpose, the Foreign Office desired to receive as detailed information as possible regarding installations planned by the United States Weather Bureau to make arrangements for the operation of this station."[38] The Danes adopted what US Ambassador in Copenhagen, Josiah Marvel, called a "mincing approach"

35 DUPI, *Grønland under den kolde krig*, p. 87.

36 Only in 1947 the US Legation in Copenhagen was raised to Embassy.

37 Dispatch No. 129, loc. cit., p. 6.

38 Ibid., p. 8.

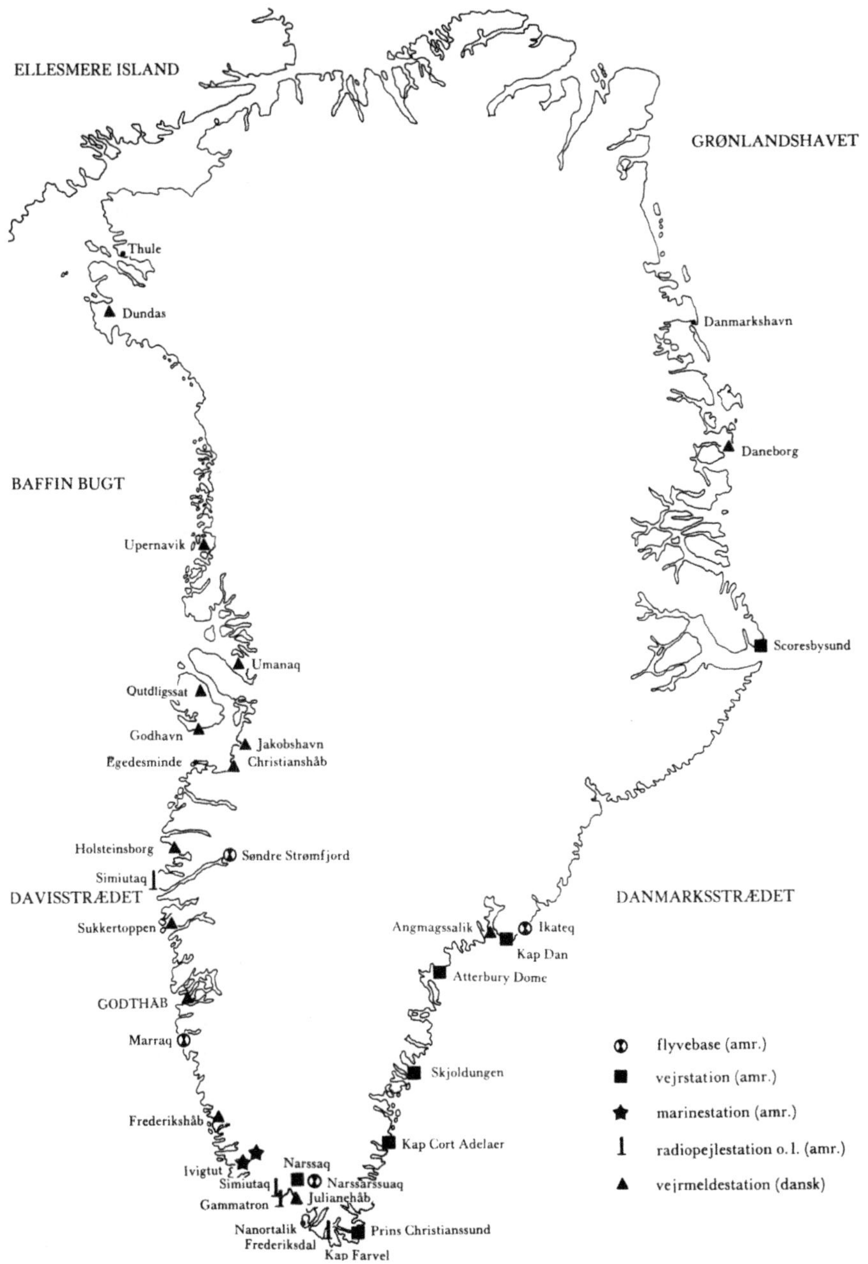

FIGURE 11.2 *War time bases and weather stations in Greenland in 1945.*
SOURCE: GADS LEKSIKON, *DEN KOLDE KRIG OG DANMARK*, (COPENHAGEN, 2011), P. 308.

to gain more time.[39] In line with the policy of the Ministry of Foreign Affairs, Denmark felt it could not decline the request and had to take over responsibility for the new station. The limited availability of staff and resources, however, rendered this policy ineffective. On 29 May, Foreign Minister Thorkil Kristensen answered the American request: Denmark supported a new weather station in Thule. As Denmark did not have the personnel at hand that would be necessary to pursue upper air observations in Thule, it would accept American personnel for the education of Danish staff in the operation of that station, before it could fully be taken over by the Danes.[40] At the same time, the Ministry of Foreign Affairs exerted pressure on the Greenland Administration and the Meteorological Institute to prevent the expansion of US activity as much as possible, by sending Danish staff. Hence, Viggo Laursen was the first to arrive in Thule and thus accepted a political rather than a scientific mission.

Negotiations about weather stations kept Denmark and the USA occupied until 1950. The new weather station in Thule was established in the summer and autumn of 1946, staffed by 11 members of the US Weather Bureau and by 11 Danes.[41] The takeover of the remaining weather stations in Greenland caused many problems in the following years, and was not completed before July 1950.[42] It took Denmark many years to provide the required staff and resources: the involved investment in weather stations was tremendous for such a small country. From 1946 until 1950, expenses for the instruments and the operation of the stations increased by a factor of four.[43] Danish Prime Minister Hans Hedtoft Hansen admitted at a press conference, in July 1948, that the operation of a weather service in Greenland required significant investments, which

39 Ibid., p. 10.

40 DUPI, *Grønland under den kolde krig*, p. 172.

41 Weather Bureau arctic meteorological service, *Weather Bureau Topics* (September 1946), p. 58; Shelagh Grant, *Polar Imperative, A history of arctic sovereignty in North America* (Vancouver, BC, 2010), pp. 297–8. The Thule weather station was the first of the Joint Arctic Weather Stations network established in 1946–1951, which comprised five more stations in the north of Canada. See Heymann, "In search of control;" Gordon W. Smith, "Weather Stations in the Canadian North and Sovereignty," *Journal of Military and Strategic Studies* 11:3 (2009), 1–63; Daniel Heidt, "Clenched in the JAWS of America? Canadian Sovereignty and the Joint Arctic Weather Stations, 1946–1972," in P. Whitney Lackenbauer (ed.), *Canadian Arctic Sovereignty and Security: Historical Perspectives* (Calgary, AB, 2011), pp. 145–69.

42 DUPI, *Grønland under den kolde krig*, p. 110. US authorities complained regularly about the poor reliability and quality of weather observations provided by Danish staff; see Heymann, "In search of control."

43 DUPI, *Grønland under den kolde krig*, p. 67.

hampered developments in other areas. But "it was unavoidable and inevitable that the weather service received support, even if at the expense of other work that was needed."[44] In the period from 1948 until 1958, Denmark's expenses for the weather service in Greenland took a 13.9 percent share of total Danish investments in Greenland, marking it as the second biggest expenditure of nine major posts (such as infrastructure, health care, factories, schools/broadcasting/churches, etc.).[45]

Sovereignty and the 'Greenland Card'

Throughout 1950 and 1951, the French anthropologist Jean Malaurie happened to live with the Inuit in Thule. He described the surprise of the Inuit upon seeing thousands of Americans coming down from the sky. After meeting with Air Force commanders suspicious of his motives, Malaurie was briefed about the otherwise top-secret Operation Blue Jay. The USA had just launched "what was said to be the biggest military enterprise since the Allied landings in Normandy." It transformed "a valley three miles wide and nine miles long, at one end terminated by a glacier, and at the other end closed off by the sea," into "the Strategic Air Command's most powerful atomic base."[46] The effort Malaurie happened to witness was the build-up of Thule Air Base in the north of Greenland. What had started with a new meteorological station in 1946 with a small airstrip added to it, turned into one of the largest US overseas bases during 1951 and 1952, and would serve as the operating base of the US polar strategy against the Soviet Union.[47] Operation Blue Jay, the construction of Thule Air Base, involved approximately 12,000 men, who built a runway of 3km in length, barracks for 5000 personnel, and other, related facilities. Thule Air Base in Northwest-Greenland became the most important hub for military operations in Greenland, as well as for scientific exploration.[48] This raises the

44 Ibid., p. 175.

45 Investments for infrastructure took a share of 24.2 percent. Axel Kjær Sørensen, *Denmark-Greenland in the twentieth century* (Copenhagen, 2007), p. 116.

46 Jean Malaurie, *The Last Great Kings of Thule* (New York, 1982), pp. 388–9, 394. Quoted after Jonathan D. Greenberg, "The Arctic in World Environmental History," *Vanderbilt Journal of Transnational Law* 42 (2012), pp. 1367–8.

47 Thule Air Base served as staging base for bombers with nuclear bombs for retaliation of a Soviet attack across the Arctic, see Petersen, "SAC."

48 DUPI, *Grønland under den kolde krig*, pp. 219–20.

question whether and, if so, why Denmark approved such a significant expansion of US military presence.

Both Denmark and the USA had an interest in stabilizing their relationship. Denmark was clearly not in a position to force the USA to leave Greenland. Likewise, the USA had to invest time and diplomacy to develop some kind of consensus regarding US military activities in Greenland. After Denmark refused to sell Greenland to the USA, negotiating a new agreement on the defense of Greenland became a priority for the US government. The negotiations began in 1947, but initial progress was slow.[49] Eventually the Danish-American Agreement on the Defense of Greenland was signed on 27 April 1951. Denmark could not refuse American access to Greenland, but negotiated strict limitations on American rights for operations in Greenland, while the USA explicitly acknowledged Danish sovereignty. The new agreement entailed the following main points: it permitted US forces to establish and operate air bases, and to pursue military activities in three defined defense areas around US bases in Narsarsuaq in the South, Søndre Strømfjord in mid-Greenland and Thule in the North of Greenland. US activities in these areas would be observed by a Danish liaison officer, to be appointed by the Danish government, who would report back to Copenhagen about US military operations. In the defense areas the USA were free to conduct scientific investigation. Research efforts outside the defense areas, however, would require advance permission by the Danish Ministry of Foreign Affairs.

While Danish sovereignty was formally respected, in practice, military sovereignty was ceded to the USA. This formed part of a silent agreement between the Danish and the US governments that was kept hidden from the Danish public.[50] The 1951 agreement allowed Denmark a significant degree of control and, most importantly, included a guarantee of sovereignty that solved the sovereignty issue once and for all, or at least that was the understanding. This was a great relief to the Danish government; leaving military control to the USA was fine by comparison. Furthermore, the Danish government interpreted concessions granted in Greenland as a contribution to NATO defense efforts—concessions granted in exchange for keeping the Danish mainland free from US Forces, thus honoring the strong neutralist and pacifist sympathies of the Danish public. The USA's interest in Greenland secretly served as an argument for the Danish government in the negotiation of advantages

49 Amstrup, "Grønland," p. 169.

50 DUPI, *Grønland under den kolde krig*, pp. 128–49; Nicolaj Petersen, "Negotiating the 1951 Greenland Defense Agreement: Theoretical and Empirical Analyses," *Scandinavian Political Studies* 21 (1998), 1–28; Petersen, "SAC," pp. 96–8.

in other areas. This became a common strategy in the following years, and Danish historians have characterized it as 'playing the Greenland card'.[51] The 1951 agreement furnished the small state with some "power of the weak" opposite the superpower.[52] But with the quick build-up of Thule Air Base, Denmark almost instantly also experienced its own limitations. Also, American research activities in Greenland grew at a pace that the Danish government had not anticipated. Thule Air Base became a hub of an expanding research activity that would eventually cover the entire island.

The 1951 agreement coincided with the Korean War and steeply rising military expenditures in the USA—expenditures which also promoted the expansion of the military and of research activities in Greenland. The largely unknown Arctic environment caused major challenges for US forces, and fueled a boom of geophysical research as a means to gain physical control of the environment, and military control of the island.[53] Close to Thule Air Base a first large research camp, Camp Tuto (shorthand for Thule take-off), was built at the edge of the ice cap, and operated from 1954 to 1966. It could house up to 1,000 men and represented a veritable research center focusing on glaciological and meteorological research (research on snow layers, snow accumulation, snow density, ice movements, crevasse studies, weather control, etc.).[54] Furthermore, Camp Tuto served as an access point to the ice cap. Further research camps were built on the ice cap, including Camp Fistclench (1957–1960) and Camp Century (1959–1966). American research activities in Greenland expanded exponentially and rapidly, so that Denmark decided to send an additional scientific officer to Thule to serve as an official advisor to the Danish liaison officer; he had the task to observe and control American research activities. This move was initiated by Danish scientists who had *de facto* lost control over US scientific activities and feared competition from American scientists. Only two years later, in 1956, the scientific advisor wrote

51 Lidegaard, *I kongens navn*, p. 763, note 379; DUPI, *Grønland under den kolde krig*, pp. 273–5; Anders Midtgaard, *Grønlandskortet: En undersøgelse af Grønlands funktion som forhandlingskort i det dansk-amerikanske forhold 1945–1961* (MA thesis, Aarhus, 2007), accessible online at www.koldkrig.dk/pdf/AM_gronlandskortet.pdf [accessed 2 April 2014].

52 Midtgaard, *Grønlandskortet*, p. 7.

53 Janet Martin-Nielsen, "The Other Cold War: the United States and Greenland's Ice Sheet Environment, 1948–1966," *Journal of Historical Geography* 38 (2011), pp. 69–80; M. Heymann, H. Knudsen, M.L. Lolck, H. Nielsen, K.H. Nielsen and C.J. Ries, "Exploring Greenland: Science and Technology in Cold War Settings," *Scientia Canadensis* 33 (2010), pp. 11–42.

54 DUPI, *Grønland under den kolde krig*, pp. 356–7; Walter Wager, *City under the Ice. Story of our Incredible Polar Base Below the Greenland Ice Cap* (Philadelphia, PA, 1962), p. 15.

in his report that it was hardly possible to gain an overview of the enormous amount of US research activities in Greenland. In 1958, a memorandum of the Danish Ministry of Foreign Affairs mentioned the same concern: "American investigations in Greenland are observed only insufficiently by the Danish authorities." On the margin of this note, the Ministry's Secretary General Axel Serup added: "From a point of view of sovereignty this is highly unsatisfactory."[55]

Camp Century: First Steps Towards the Total Militarization of Greenland

One of the most striking research efforts in Greenland opens up another perspective on US-Danish relations in Greenland: Camp Century. Many of the American research efforts represented preparations for the construction of a 'city under the ice,' i.e., Camp Century. The construction of Camp Century commenced in June 1959, with convoys of large Caterpillar D-8 and D-9 tractors. Within two years, these transported 5,600 tons of materials over the ice cap, 225 km from Thule Air Base to the East. In about 15 months a system of 21 tunnels was built under the ice. The tunnels contained a small city complete with a hospital, shops, theater, and a church. Camp Century accommodated up to 225 inhabitants and served mainly for basic research efforts in fields like glaciology, Arctic meteorology and others. From 1961 until 1963, electricity for Camp Century was provided by means of the world's first portable 1.5 Megawatt nuclear reactor 'Alco PM-2A.' Additionally, three Diesel generators were installed as an emergency supply; they contained Diesel fuel that would last for 190 days.[56] Camp Century was the testing ground for an ambitious plan of the US Army: Project Iceworm. The Army pursued the idea of storing 600 so-called 'Icemen,' i.e., Medium Range Ballistic Missiles, under the Greenland ice cap, hidden in tunnels and distributed over an area of 135,000 square kilometers (an area equivalent to the size of Greece). The Army boasted of significant

55 Danish Ministry of Foreign Affairs, "Den Videnskabelige Rådgivers (VR) afsluttende rapport for 1956," 16 June 1958, DNA, Copenhagen, UM 105.F.8, Videnskabelig Rådgiver for FOTAB.

56 Henry Nielsen and Kristian H. Nielsen, "Camp Century—Cold War City under the Ice," (manuscript, Aarhus, 2013), forthcoming in: Matthias Heymann, Kristine C. Harper (eds.), *Exploring Greenland: Expanding Technologies and Creating Geophysical Knowledge in the Cold War* (New York: Palgrave Macmillan); Kristian H. Nielsen, Henry Nielsen and Janet Martin-Nielsen, "City under the Ice: The Closed World of Camp Century in Cold War Culture," *Science as Culture* 23 (2014), 443–464; Wager, *City under the ice*; Charles Michael Daugherty, *City under the Ice—The Story of Camp Century* (New York, 1963).

FIGURE 11.3 *Transport of a nuclear reactor into the tunnel system of Camp Century.*
SOURCE: CHARLES MICHAEL DAUGHERTY, *CITY UNDER THE ICE—THE STORY OF CAMP CENTURY* (NEW YORK, 1963), P. 118.

advantages of this system: the ice shield was remote from any large industrial and residential centers; and the missiles were hidden and could periodically be moved to reduce vulnerability in the case of a nuclear war. With this plan the US Army hoped to gain ground in its fierce inter-service competition with the Air Force and Navy, both of which already commanded nuclear forces at the time. While Camp Century was built successfully, Project Iceworm was stopped by Secretary of Defense McNamara in 1963.[57]

The American intent to build Camp Century was certainly an affront to the Danish authorities and likely to cause problems if the plans were made public.[58] On November 4, 1958, the US Embassy contacted the Danish Ministry of Foreign Affairs to convey an informal message: the US Army would be seeking Denmark's permission to build another scientific camp, i.e., Camp Century, on the Greenland ice cap. The new camp would be extraordinary in comparison with the previous ones: it would be nuclear-powered and located entirely inside the ice cap. After internal discussions in the Ministry, a memorandum to the Americans, dated 19 November, recommended that the USA either give up the idea or build the camp outside of Greenland—in Alaska or Northern Canada. The memorandum stated: "We have come to the conclusion that experiments with nuclear reactors in Greenland will give rise to a number of problems that we rather [...] would like to avoid."[59] The Danish Government feared two things: firstly, sharp reactions and pressure from the Soviet Union[60] and secondly, harsh reactions from the Danish media and the Danish public. Even though a nuclear reactor was not a nuclear weapon, it was still likely to cause much controversy, especially since the ban of nuclear weapons had just been enacted in Denmark.

On 29 May 1959, ignoring the Danish response, the US Embassy in Copenhagen presented a formal request to the Ministry of Foreign Affairs for permission to construct a nuclear-powered camp under the ice cap. Three months later, on 18 August 1959, US ambassador Val Petersen met Danish Foreign Minister (and

57 Nikolaj Petersen, "The Iceman That Never Came: 'Project Iceworm', the Search for a NATO Deterrent, and Denmark, 1960–1962," *Scandinavian Journal of History* 33 (2008), pp. 75–98; Erik D. Weiss, "Cold War under the Ice: The Army's Bid for a Long-Range Nuclear Role," *Journal of Cold War Studies* 3 (2001), pp. 31–58.

58 The following part of the story is investigated in detail in Nielsen and Nielsen, "Camp Century."

59 Danish Ministry of Foreign Affairs, "Note," signed Axel Serup, 19 November 1958, Danish National Archive (hereafter DNA), Copenhagen, UM 105.F.2b/2.

60 In March 1957, the Soviet Union had reacted sharply to Denmark's decision to acquire Honest John and Nike missiles to strengthen its air defense; Nielsen and Nielsen, "Camp Century," p. 22.

later Prime Minister) Jens Otto Krag at a cocktail party and informed him that the US Army had already begun to build the research camp; that the Army had been pressed for time and had not received any answer to its formal request from the Danish government. Krag was shocked. He considered the American activity a serious breach of contract. He instantly called leading representatives of the Danish administration for an 'emergency meeting' that was to take place two days later. All of its attendees shared Krag's concern. Hans Henrik Koch, head of the Danish Atomic Energy Commission's executive committee, suspected that "there must be other purposes for the nuclear reactor than the ones given by the Americans[;] [...] he [...] worried that the installation of the reactor served also military purposes."[61] Military science, however, was an issue that still heavily conflicted with the ideology of science as a peaceful endeavor, which still was prevalent in Denmark. An official from the Ministry of Foreign Affairs feared "that the press would raise the issue, addressing the very question of nuclear weapons in Greenland."[62] On 26 August, the Danish Ministry of Foreign Affairs indicated the possibility of a positive answer to the US embassy, but demanded more information and full secrecy about these plans. It particularly asked the USA to refrain from informing the press about the project, "as such indiscretion would create great difficulties for the Danish authorities."[63]

But the Danish request for discretion came too late. The American newspaper *The Sunday Star* had published an article entitled "City Under Snow in Greenland To House Army Post" three days earlier, including detailed descriptions of plans for the nuclear reactor. On 7 September 1959, in lack of feasible alternative options, Krag briefly advised the Foreign Affairs Committee of the Danish Parliament, that "we have received from the American Government a request for permission to install for civil purposes a nuclear reactor in the vicinity of Thule. The government did not have reservations about responding in a positive manner to the request."[64] The Danish government, unable to keep Camp Century out of the public eye, reverted to a different strategy instead, in order to avoid upheaval in the domestic media: it tried to convince the Parliament and the Danish public that the project was entirely civilian,

61 Quoted in Danish Ministry of Foreign Affairs, "Minutes of the meeting," 20 August 1959, p. 12, DNA, Copenhagen, UM 105.F.2b/2.

62 Ibid.

63 Danish Ministry of Foreign Affairs to US Embassy, 26 August 1959, DNA, Copenhagen, UM 105.F.2b/2.

64 Danish Ministry of Foreign Affairs, "Minutes of the meeting," 7 September 1959, DNA, Copenhagen, UM 105.F.2b/2.

not military, in nature.[65] Additionally, the Danish government tried to control the release of information pertaining to the facility, particularly its strategic significance and the presence of a nuclear reactor. This strategy proved to be untenable, especially after the US Army launched a publicity campaign about the unusual project. Many journalists and one television crew, led by prominent CBS reporter Walter Cronkite, visited the city under ice. The US Army also produced a 30-minute feature about Camp Century, which was broadcast in 1961 as part of the Army's "The Big Picture" series on ABC-TV.[66]

With this publicity, knowledge about the military significance of Camp Century was circulated, although the existence and aims of Project Iceworm remained secret until revealed by historians of the 1980s and 1990s. In the much-circulated popular journal *Reader's Digest*, American readers learned as early as in May 1960 (and readers of the Danish version in June) about Camp Century that "working areas will be dominated by scientific laboratories, but should the need come for a similar military installation—perhaps an under-ice launching site for intercontinental ballistic missiles or interceptors—we have the blueprints, the techniques, the machines."[67] The Danish resistance to the construction of Camp Century and—later, when this strategy proved untenable—Danish attempts to contain information of the kind found in the *Reader's Digest* article, may be seen as first signs of a new Danish stance on US military activities in Greenland. Partly due to the secret accord between Denmark and the USA, enacted by H.C. Hansen's letter, the question concerning nuclear facilities, even nuclear reactors (as in the case of Camp Century), had become a highly problematic issue for the Danish Government, which feared a public controversy. Leading politicians and civil servants knew that something had to be done, although they were unsure what, and how. The Thule crash forced the Danish authorities to confront the 'nuclear issue' head on.

---

65 Nielsen and Nielsen, "Camp Century," p. 13.

66 Ibid., p. 20. The Danish government still succeeded in censoring important publications about Camp Century. For example, it successfully requested the elimination of passages referring to the military character of Camp Century from Walter Wager's popular book; ibid., pp. 20–23. On publicity about Camp Century: see Nielsen et al., "City under the Ice." On publicity campaigns of the Army see D.J. Kinney, "Selling Greenland: The Big Picture Television Series and the Army's Bid for Relevance during the early Cold War," *Centaurus* 55 (2013), pp. 344–57.

67 Herbert O. Johansen, "City Under Ice," *Reader's Digest* (May 1960), pp. 176–9; Danish version: "Det Bedste," (June 1960), quoted in DUPI, *Grønland under den kolde krig*, p. 325. See also Petersen, "The Iceman," p. 78.

Operation PCA 68: New Foundations of the US-Danish Relations

The 1960s saw significant changes in the strategic role of Greenland. The development and deployment of long-range missiles on both sides of the iron curtain reduced the role of Thule Air Base (AB) as an offensive base for bombers equipped with nuclear weapons. Conducting reconnaissance flights and providing early warnings of missile attacks became the main mission for the Air Base. In 1958, the US government decided to build the Ballistic Missile Early Warning System (BMEWS), consisting of three large radar stations in the high North in Thule, Greenland, in Clear, Alaska, and in Fylingdales Moor, England. Upon completion of the BMEWS station in 1960, Thule AB was handed over from Strategic Air Command to Air Defense Command.[68] This change reflected a shift in the strategic role of Thule AB and affected military and scientific operations in Greenland drastically. Both Camp Tuto and Camp Century were closed in 1966. The focus of US military research interests shifted from glaciological research on the ice cap in the 1950s to communication-related atmospheric and ionospheric research, which became the predominant scientific activity in the late 1960s.[69]

The frequent occurrence of Northern Lights in the Arctic disrupted radio communication and hampered military operations. Hence, the investigation of atmospheric disturbances in the ionosphere received increased attention. Ionospheric research became one of the few research fields in which Danish scientists were heavily involved; they also profited from ample US military funding for this purpose. In 1967, a particularly ambitious research plan, named PCA 68, was suggested to the Danish government.[70] On 21 November, the US

---

68 Little has been written about BMEWS. Sagan, *Limits*, pp. 162–3; Petersen, "SAC," pp. 110–2; Olesen, "Tango," pp. 121–22; Kenneth Schaffel, *The Emerging Shield: The Air Force and the Evolution of Continental Air Defense, 1945–1960* (Washington DC, 1990), p. 259. On the British part of BMEWS see Jeremy Stocker, *Britain and Ballistic Missile Defence 1942–2002* (London, 2004), pp. 81–98. For a technical account, see Melvin L. Stone and Gerald P. Banner, "Radars for the Detection and Tracking of Ballistic Missiles, Satellites, and Planets," *Lincoln Laboratory Journal* 12 (2000), pp. 217–44.

69 Henrik Knudsen, "Battling the Aurora Borealis: The Transnational Coproduction of Ionospheric Research in Early Cold War Greenland," (manuscript, Aarhus, 2013), forthcoming in: Matthias Heymann, Kristine C. Harper (eds.), *Exploring Greenland: Expanding Technologies and Creating Geophysical Knowledge in the Cold War* (New York: Palgrave Macmillan).

70 This account is based on Henrik Knudsen, "Cold War Militarization of Space, Ionospheric Research in Greenland, and the Politics of Rockets. A Case Study of the Ill Fated Operation PCA 68" (unpublished working paper, Centre for Science Studies, Aarhus University, 2011).

Embassy sent a confidential note, along with a number of technical reports, to the Danish administration which presented an outline of two rocket projects to be carried out by the US Air Force and Defense Atomic Support Agency at Thule Air Base during the autumn of 1968. The comprehensive operation conducted by the Air Force Cambridge Research Laboratory included the plan to launch 37 rockets from a "refurbished Nike-Hercules Rising Star Complex" near Dundas Mountain to the north of Thule AB.[71] As a second part of the project, Danish ionospheric researchers from the Technical University of Denmark and the Danish Meteorological Institute planned to conduct a smaller program; this was to launch 10 rockets of a smaller size, in cooperation with the American scientists. Moreover, the project included a coordinated series of ground-based, satellite and airborne measurements. Some 100 scientists and support personnel were to work on this project, which boasted a total budget exceeding 10 million dollars.

This project served to investigate a period of maximal solar activity in the sun spot cycle, which occurred in autumn 1968. A strong influx of high energy protons would ionize the lower part of the atmosphere and heavily disrupt radio communication. The motivation behind the project was not purely scientific, but military in nature. Since the Limited Test Ban Treaty of 1963, the direct investigation of effects of an atmospheric nuclear detonation on communication and communication facilities had become practically impossible. Operation PCA 68 promised to provide a valuable alternative. An internal explanation by the Department of Defense defined the goals as follows: "To better understand nuclear weapons effects in the upper atmosphere more complete knowledge is needed of the atmospheric chemistry, coupling mechanisms between the expanding fireball and the ambient air, causes of debris distribution and high altitude plasma-magnetics. Certain natural disturbances offer an ideal opportunity to gain information about the physical chemistry in the disturbed lower ionosphere [...]. These data will be used to develop a model of the disturbed ionosphere. Extrapolation can be made to higher energy levels to develop a more accurate model of the nuclear fireball and the highly disturbed ionosphere."[72] The impact area of the rockets for investigating atmospheric disturbances was defined as an area on the Baffin Bay, just West of Thule.

71 J.C. Ulwick, *Operation PCA 68* (Air Force Cambridge Research Laboratories, 1968), Ministry of Greenland, Departementschefs Archive (confidential), DNA, Journalsager 1950–78, XLI B V, PCA Raket Program (hereafter DNA, PCA Raket Program).

72 Office of Aerospace Research, *Air Force Research Resumes 1968*, p. 484.

Contrary to most US research activities in Greenland since the 1950s, Danish scientists wanted to get involved in Operation PCA 68.[73] The Danish collaborator from the Ionospheric Laboratory at the Danish Technical University, Jens Olesen, assured his American colleagues in December 1967 that Danish researchers would do whatever was in their power to make the application "go fast through the official channels here."[74] The Danish government reviewed the application with great care. Ship and air traffic needed to be safeguarded, a safety zone defined and the consequences for the local Greenlandic population investigated. On May 22, 1968, the Ministry of Foreign Affairs received a response from the local administration, explaining that "lively traffic" took place in the area during the months in question, "since the hunt for whales among other things takes place from small vessels" in the area in question. Cordoning off the area would "absolutely have economic consequences for the local population," since hunting activity in this area, and at this time of the year, were critical for the successful procurement of winter supplies. Further, strong protests were raised by local members of the national council of Greenland.[75] On May 27, leading members of the Danish Ionospheric Laboratory were briefed at the Ministry of Greenland. Jens Olesen voiced strong concerns about the consequences of a last-minute rejection "on a meager basis such as the present case." He held that "[a] rejection will reverberate in all parts of the international circle of scientific groups within space science, it will seriously harm the reputation of Denmark in general and in particular concerning future cooperation with the U.S.A., especially given the motivation for the rejection, i.e., exclusion orders which would ban a very small number of North Greenlandic hunters from a certain sea area for in total less than 36 hours during two months."[76]

73 Ionospheric research is one of the few research fields with established collaborations between Danish and US scientists since the 1950s. Another example is the work of Danish ice core researcher Willi Dansgaard in the 1960s: Knudsen, "Battling the aurora;" Janet Martin-Nielsen, "'The Deepest and Most Rewarding Hole Ever Drilled': Ice Cores and the Cold War in Greenland," *Annals of Science* 70 (2013), pp. 47–70; Maiken Lolck, *Klima, kold krig og iskerner* (Aarhus, 2006).

74 J.K. Olesen to James C. Ulwick. 19 December 1967, DNA, PCA Raket Program.

75 City treasurer [on behalf of the inspector] to the Ministry of Greenland, 22 May 1968, DNA, PCA Raket Program.

76 J.K. Olesen to Karl Andersen, Director of the Danish Meteorological Institute, 28 May 1968, DNA, PCA Raket Program.

The Danish government considered financial compensation for the small local Greenlandic population, to buy its way out of the dilemma. It also asked additional Danish scientists who were familiar with the area for further investigations. In the final, detailed presentation to the Danish Foreign Minister on 14 June 1968, the project was recommended by the responsible officers, after a detailed discussion of safety issues, the assessments of local authorities, and the evaluation of Danish scientific interests. Due to the military character of the project the topic at hand was controversial. The proponents countered that, firstly, all American projects in Greenland were "military scientific activities" and, secondly, that Denmark had a strong "interest that western defensive radar and communication systems operate efficiently."[77] On 27 June, after repeated pressure from the US authorities who had planned to ship materials to Thule, Prime Minister Hilmar Baunsgaard, the Minister of Foreign Affairs Poul Hartling and the Minister of Greenland A.C. Normann decided to suspend the program and make an allowance for reconsiderations in 1969. The political rationale, documented in a short note, was the wish to avoid "any possible new uneasiness about Thule *even* if it appears unmistakably that such uneasiness isn't *factually* justified." A note to the US Embassy followed on 1 July, and an explanation four days later.[78] The Danish scientists either involved in or otherwise supporting the planned project were shocked. The Director of the Danish Geodetic Survey, Einar Andersen, informed Niels Boel at the Ministry of Foreign Affairs that the decision was "unwise" in his opinion, and that he "as a Danish scientist felt ashamed." "Our American friends are of course quite different from us, and we often become rather annoyed by them, but we must acknowledge that they have done a very great deal for us and made a great many things possible here."[79] It seems that by the late 1960s, Danish scientists, unlike Danish politicians, had grown to accept the hegemonic presence of the US in Greenland, and had grown accustomed to utilizing this presence in gaining support for their own goals.

---

77 Danish Ministry of Foreign Affairs, "Note," signed by Gunnar Blæhr (14 June) and Niels Boel (19 June), 1968, DNA, Copenhagen, UM, 105.F.9.a.

78 Danish Ministry of Foreign Affairs to Danish Embassy in Washington, 4 June 1968, DNA, Copenhagen, UM 105.F.2.a; emphasis added in the original.

79 Einer Andersen to Niels Boel, 17 July 1968, Danish Ministry of Foreign Affairs, DNA, Copenhagen, UM, 105.F.9.a.

Conclusion

The strategically important location of the Danish colony of Greenland (which, since 1953, was an autonomous province), and the USA's strong military interest in it, put Denmark under significant pressure. The Danish government never considered selling Greenland to the USA, and thus did not take up the US government's preferred option. But it had to compromise its interest in various ways in order to regain full sovereignty and control of Greenland after the liberation of Denmark in 1945. The USA respected Danish sovereignty in principle, but did not hesitate to put the Danish government under strong pressure with numerous requests, often phrased with determination and handed over at short notice. The Danish government often felt pressured to give in to US demands, as was the case for the permission to run a nuclear reactor at Camp Century, or indeed to fulfill American requests, as in the operation of the full number of weather stations. Otherwise, Denmark feared that the USA would simply start to implement its plans without official Danish consent. By contrast, the USA attempted to respect Danish interests and sensitivities only within certain limits. In practice, the USA curtailed Danish sovereignty, while the Danish government managed to avoid damaging domestic debates.

The USA soon expanded scientific and military activities on the island, controlled Greenland from a military perspective from the outset, and assumed the defense of the island. This American hegemony was certainly not consensual but forced upon the small state.[80] Denmark's decision to join NATO in 1949 did result in a purely consensual and friendly relationship. Denmark remained an ally with clear reservations.[81] Sovereignty in Greenland was a main concern, and further pressures accumulated. Situated between two hostile superpowers, Denmark feared Soviet hostility for any advantage that it granted to the USA. Furthermore, the Danish government (irrespective of which party was ruling) was concerned about domestic public unrest because of developments in Greenland, particularly given the strong neutralist and pacifist as well as nationalist traditions and sympathies in the country.[82]

---

80 John Krige, *American Hegemony and the Postwar Reconstruction of Science in Europe* (Cambridge, MA, 2006). See also Heymann, Nielsen, "Introduction," pp. 230–4.

81 Poul Villaume, *Allieret med forbehold—Danmark, NATO og den kolde krig, En studie i dansk sikkerhedspolitik 1949–1961* (Copenhagen, 1995).

82 Carsten Holbrath, *Danish Neutrality: A Study in the Foreign Policy of a Small State* (Oxford, 1991); Hans Branner and Morten Kelstrup (eds.), *Denmarks Policy Towards Europe after 1945: History, Theory and Options* (Odense, 2003).

In the immediate post-war years, Denmark was in a difficult situation and thus unable to withstand American demands. The Danish government hoped to avoid bilateral negotiations with the USA in preference of multilateral negotiations. This strategy usually failed, as for Greenland's weather stations. While the Provisional International Civil Aviation Organization demanded the operation of 14 stations, the Danish Ministry of Foreign Affairs felt forced to give in to the American demand of 24 stations for weather observation. In order to keep the US forces out of Greenland as far as possible, the Danish government attempted to cover all tasks by Danish staff and replace American staff, even in the case of the requested new weather station in Thule. This policy was unrealistic, given the limited resources available to Denmark. The Danish government also repeatedly delayed actions and sent unclear or contradictory messages to US representatives, or indeed no messages at all. When the US Coast and Geodetic Survey asked for research results of the Danish Geodetic Institute in June of 1946, it did not receive the requested information. "One can say that it was characteristic for Danish politics in this question, that after detailed deliberations the decision was chosen, not to respond to the request," historian Niels Amstrup concluded.[83]

Danish concerns were relieved to some extent after the 1951 agreement was signed, which put Danish-US relations in Greenland on firmer ground. While Denmark was satisfied with US guarantees and the control it achieved, the agreement was traded for silent consent. The negotiations about weather stations alone showed ambivalence in Danish policy with regard to the Danish public. Danish rights and ambitions to defend full sovereignty were clearly articulated, but Danish concessions were hidden from the public to the highest possible extent. The Ministry of Foreign Affairs refused to sign a contract about weather stations in June 1946 in Copenhagen. Instead, it asked the American Legation to send a request to District Governor Simony in Greenland, its softened prose avoiding military connotations or the topic of the negotiations in Copenhagen. This careful maneuvering was the beginning of a tradition of two-edged policies, which intensified during the 1950s and led to a stronger discrepancy between open and hidden agendas. The military role of Greenland was publicly deemphasized, military activities represented as civilian and the cessation of military sovereignty to the USA masked by explicit sovereignty guarantees from the USA. Nuclear bombs in or over Greenland remained hidden behind a veil of deliberate ignorance. Soon, this policy even required censorship of the press, for example in relation to Camp Century. Foreign Minister Jens Otto Krag, trapped in a self-inflicted duplicitous policy, could do nothing

83 Amstrup, "Grønland," p. 167.

other than accept the American affront and filter information for the public when he learned about the creation of the Camp. It is not clear whether a fear of Russian threats or of domestic scandal was the more important motivation for him. This policy came at a cost, but 'playing the Greenland card' also had its advantages.

It took a scandal to cut through this Gordian knot of international relations, and this scandal presented itself in the Thule crash of January 1968. Krag succeeded in terminating the policy of duplicity with regard to nuclear weapons and Greenland in an elegant manner. (At the same time, he kept the scandal hidden from the Danish public, thereby continuing the strategy of double-dealing to a certain extent.) While the crash certainly shocked the Danish government, it allowed Krag to gain new ground for Denmark in negotiations with the USA. This time, Krag could safely embarrass and defy the US government to free Denmark from its double bind. Historian Thorsten Borring Olesen has concluded that the Thule crash was a "blessing in disguise" for the Danish government.[84] This time, the USA was forced to respond on terms dictated by the Danish and, as a consequence, had to redefine its policy in Greenland fundamentally. The case of Operation PCA 68 indicates that a new era had begun in 1968, in which the Danish Government learned to better withstand American pressure, and rated potential domestic concerns higher than expected scientific gains of a rocket project or diplomatic irritations—even when faced with internal pressure from domestic scientific elites with vested interests in American infrastructure, funding and collaboration. This was not the first time that the Danish government refused permission to a scientific enterprise, most of which had been granted in the past. Stopping the extraordinarily ambitious and well-prepared undertaking PCA 68, which even Danish scientists backed with confidence, represented a powerful signal that Denmark was no longer prepared to always bow to US interests.

84 Olesen, "Tango," p. 147.

CHAPTER 12

The Ford Foundation and the Measurement of Values

Paul Erickson

In May of 1954, an article titled "An Approximate Measure of Value" appeared in the *Journal of the Operations Research Society of America*. The article's subject matter appears, at first glance, rather removed from the concerns of operations research and 'management science,' new quasi-disciplines that emerged in the wake of World War II. It certainly stands out among the other articles appearing in the same volume, which include an investigation of traffic delays at the New York Port Authority tolls, and an application of optimization theory to the scheduling of military fuel tanker deliveries. With their article flanked by these seemingly mundane inquiries, the authors sought to "give a method of estimating the values an individual associates with outcomes [...] that seems amenable to practical application." And while they would admit that their method was based on assumptions that substantially restricted its applicability, they nevertheless noted that their method had already achieved application to problems of quality control in the Commercial Solvents Corporation, and the five-year planning process of the Barth Corporation of Cleveland.[1]

Read from the perspective of the present, the article seems incongruous. But wittingly or not, its authors had stumbled upon what was perhaps *the* pivotal question of their technocratic age, which would re-emerge as the focus of numerous books, articles, and conferences between the 1940s and the 1960s that blended the exalted and the mundane, the philosophical and the practical: how could 'values' be discerned from the inaccessible minds of human beings, measured, and used to formulate policy? It is possible to sketch many distinct histories of this question. One runs through psychophysical debates about the validity of attempts to measure sensation and to square the measurement of subjective quantities with the imperatives of behaviorism. Much of the history of utility theory in economics was marked by similar concerns. Could utility be measured? And—crucially for those concerned with

1 C. West Churchman and Russell L. Ackoff, "An Approximate Measure of Value," *Journal of the Operations Research Society of America* 2 (1954), 172–87, p. 173.

 | DOI 10.1163/9789004264229_013

the welfare effects of state intervention in the economy—could it be compared and transferred between individuals? While measurable utility theory had repeatedly been declared defunct, new proposals and frameworks for treating the concept emerged with some regularity. The topic of 'values' also had long been a major focus of discussion in the American social sciences, as part of a set of disciplinary debates about the objectivity of the social scientist, the proper relationship between social research and social policy, and the challenge of moral relativism, especially in light of the rise of totalitarian regimes in Europe.[2]

However, operations researchers and social and behavioral scientists of the 1940s and 1950s seem less inclined toward philosophical purism than their pre-war predecessors, and perhaps more confident in their technical wizardry. Rather than banish subjective quantities from science and take refuge in indifference curve analysis or the austere epistemological perspectives of behaviorism or operationalism, they would turn to a stern theoretical and methodological rigor in their attempts to peer inside the minds of others. New ideas emerging in a wide range of disciplines all speak to this trend: latent structure analysis, axiomatic theories of measurement (psychological and otherwise), an ongoing enthusiasm for psychological testing, and finally, designs for machines (real or imaginary) that would give behavioral scientists access to their subjects' thoughts and emotions.[3] As new streams of patronage drew social and behavioral scientists further into complicity with the aims and projects of the national security state, the need to get a more trustworthy fix on the values and beliefs of their subjects only grew in importance.

This essay therefore focuses on understanding the brief post-war vogue for surveying (and especially for *measuring*) values, by examining some of the early activities leading to the creation of the Ford Foundation's program in Individual Behavior and Human Relations (later renamed the 'Behavioral

---

2 The classic work in this vein is Edward A. Purcell, *The Crisis of Democratic Theory: Scientific Naturalism and the Problem of Value* (Lexington, KY, 1973). See also David Hollinger, *Science, Jews, and Secular Culture: Studies in Mid-Twentieth-Century American Intellectual History* (Princeton, NJ, 1996), especially chapters 5–8; Mark C. Smith, *Social Science in the Crucible: The American Debate Over Objectivity and Purpose, 1918–1941* (Durham, NC, 1994); Peter Novick, *That Noble Dream: The "Objectivity Question" and the American Historical Profession* (Cambridge, 1988).

3 On these impulses among post-war social and behavioral scientists, see especially Rebecca Lemov, "'Hypothetical Machines': The Science Fiction Dreams of Cold War Social Science," *Isis* 101 (2010), 401–11.

Sciences Division').[4] The essential timeline of the Foundation's creation is well known. After becoming one of the world's wealthiest philanthropic organizations after the death of Henry Ford in 1947, the Foundation, through its officers, articulated an agenda that reflected perfectly the mindset of liberal elites at the time: a firm commitment to the potential and promise of science and technology to solve even society's most persistent problems such as hunger and poverty, class antagonism, and international conflict, to name but a few. In particular, the Foundation's administrators devoted a portion of their resources to supporting basic research into human behavior, motives, and needs, on both individual and collective levels. Their rationale was that the knowledge produced in this process would create a "rational basis for planning and responsible decision-making" by disinterested, technocratic elites that was nevertheless free and democratic.

From the outset the Foundation's administrators recognized the difficulty of incorporating 'values' into their philanthropic efforts. How could the values of individuals, on whose behalf the foundation purported to work, be incorporated into their "planning" and "decision-making?" Indeed, the question arose in a very practical context during the initial meetings of the Committee on Policy and Program, which set the agenda for the Foundation as a whole. The peculiar way in which the Committee answered this question reflects, on a microcosmic level, some of the inclinations, tendencies, or impulses that would permeate similar attempts to measure values and gain access to the inner worlds of human subjects throughout the post-war era. This was in part because the Foundation funded much behavioral science research in the early 1950s, and in part because of the way its leadership had its fingers on the pulse of the age. Specifically, their approach to values steered away from a disciplinary model of specialized expertise, and toward a cross-disciplinary approach reliant on theoretical and methodological rigor.[5]

---

4 The vogue for values was, indeed, brief; by the 1960s it was receding in many parts of the social sciences. See especially the literature review in Steven Hitlin and Jane Allyn Piliavin, "Values: Reviving a Dormant Concept," *Annual Review of Sociology* 30 (2004), 359–93.

5 A number of recent works have begun to explore the 'human sciences' as a whole rather than focusing on the history of particular disciplines; see e.g., Hunter Crowther-Heyck, "Patrons of the Revolution: Ideals and Institutions in Postwar Behavioral Science," *Isis* 97 (2006), 420–46; Jamie Cohen-Cole, "Instituting the Science of Mind: Intellectual Economies and Disciplinary Exchange at Harvard's Center for Cognitive Studies," *The British Journal for the History of Science* 40 (2007), 567–97; Joel Isaac, *Working Knowledge: Making the Human Sciences From Parsons to Kuhn* (Cambridge, MA, 2012); and Paul Erickson, Judy L. Klein, Lorraine Daston, Rebecca Lemov, Thomas Sturm and Michael D. Gordin, *How Reason Almost Lost Its Mind: The Strange Career of Cold War Rationality* (Chicago, 2013).

Theory, Practice, and the Historiography of the Cold War Human Sciences

The emergence of the Ford Foundation's concern with the 'measurement of values' has interesting historiographical implications because it suggests links between several areas of scholarship on the intellectual history of the Cold War era that have remained relatively separate to date. As Philip Mirowski, S.M. Amadae, and Hunter Heyck (among others) have shown, a key focus of theory development in the post-war social and behavioral sciences were formal models of rational decision-making. Game theory, social choice theory, optimization theory, and utility theory are the prime examples.[6] Histories of these models have typically treated them as a product of a postulatory-deductive enterprise; and if one considers the essentially mathematical nature of the work of individuals like John von Neumann, John Nash, Kenneth Arrow, or John Harsanyi, it is difficult not to construct such a history. Moreover, it is tempting to read the work of these individuals primarily as an exercise in building depictions or representations of individual behavior and social interaction, or at least as explorations of logically possible systems of social organization.

Such histories exist alongside others—for example, Sarah Igo's *The Averaged American* or Ellen Herman's *The Romance of American Psychology*—that touch on the polling and survey research that blossomed in the post-war era. These accounts follow a much more empirical kind of social science that aimed to chart Americans' attitudes toward some of the great issues of the day: sexual orientation, politics, and racial attitudes, among others. Yet they potentially share territory with the literature on histories of rational choice mathematics. Consider, for example, Amadae's *Rationalizing Capitalist Democracy*: not only did liberal economists and political theorists seek to demonstrate that stable and welfare-enhancing social orders could form from the interactions of rational and free individuals, but they sought to critically examine the notion that some superintending state authority could claim to act on behalf of 'the common good' or a collective sentiment. This latter agenda spoke directly to work of sociologists and political scientists using polls, surveys, and structured interviews to shed light on the attitudes, opinions, and beliefs of communities or nations, inasmuch as they sought to use the resulting information as a guide

6 Hunter Heyck, "Producing Reason," in Mark Solovey and Hamilton Cravens (eds.), *Cold War Social Science: Knowledge Production, Liberal Democracy, and Human Nature* (New York, 2012), pp. 99–116; Sonja M. Amadae, *Rationalizing Capitalist Democracy: The Cold War Origins of Rational Choice Liberalism* (Chicago, 2003); Philip Mirowski, *Machine Dreams: Economics Becomes a Cyborg Science* (Cambridge, 2002).

to policy. But where the former group might have seen their task as one of theorizing or 'modeling' social processes of decision-making, buying and selling, or voting, the latter would have conceptualized their work as an exercise in empirical social science.

Bridging these accounts requires a richer conception of the role of 'theories' in the human sciences, beyond commonplace references to their explanatory power or predictive capabilities. A number of works, appearing over the past decade, are particularly inspiring in this regard. Historians of physical sciences, like Andrew Warwick and David Kaiser, have suggested we interpret theorizing as a form of work or practice (analogous to experimental or observational practice) involving "theoretical tools," particular notations or calculational techniques that are selectively deployed to accomplish particular ends. In addition, studies of the transformative effect of new notational systems and "paper tools," or of the use of models and simulations, successfully blur the line between the activity of theorizing and other forms of work performed in the sciences, such as experimentation or data collection.[7] Other studies have shown how techniques and working practices of data analysis in the sciences can become transmuted into 'theory,' in the sense of statements about the nature of things. Thus, as Gerd Gigerenzer has argued, the introduction into psychology of tools from inferential statistics led to a new depiction of the human mind as an "intuitive statistician."[8] Joel Isaac has likewise called attention to what he identifies as a post-war concern with "epistemic design" in the human sciences, i.e., with the notion that new theories and "conceptual schemes" might emerge from the "right arrangement of the empirical findings of research."[9] But conversely, it also seems plausible that theories, even ones

7 See especially David Kaiser, *Drawing Theories Apart: The Dispersion of Feynman Diagrams in Postwar Physics* (Chicago, 2005) and Andrew Warwick, *Masters of Theory: The Pursuit of Mathematical Physics in Victorian Cambridge* (Cambridge, 2003); on "paper tools" see Ursula Klein, *Experiments, Models, Paper Tools: Cultures of Organic Chemistry in the Nineteenth Century* (Stanford, CA, 2003); on 'models' and their significance across the natural and social sciences, see Angela Creager, M. Norton Wise and Elizabeth Lunbeck (eds.), *Science Without Laws: Model Systems, Cases, Exemplary Narratives* (Durham, NC, 2007); Mary S. Morgan, *The World in the Model: How Economists Work and Think* (New York, 2012).

8 Gerd Gigerenzer, "Where Do New Ideas Come From? A Heuristic of Discovery in the Cognitive Sciences," in Maria C. Galavotti (ed.), *Observation and Experiment in the Natural and Social Sciences* (Dordrecht, 2003), pp. 99–139; Gerd Gigerenzer, "From Tools to Theories: A Heuristic of Discovery in Cognitive Psychology," *Psychological Review* 98 (1991), 254–67.

9 Joel Isaac, "Epistemic Design: Theory and Data in Harvard's Department of Social Relations," in Mark Solovey and Hamilton Cravens (eds.), *Cold War Social Science:*

that are products of a mathematical imagination operating at some remove from empirical study, might suggest productive categories, terminological distinctions, accounting schemes, choices of independent and dependent variables, and so forth, for structuring the production of knowledge. Theory, in short, might provide not only sets of 'predictions' or 'regularities' that the phenomena under study would follow, but tools and methodologies for social and behavioral scientists themselves to make new, or newly discovered, phenomena visible.[10]

The question remains as to why some sets of tools for the production of knowledge are selected as opposed to others. It is possible to tell many different stories about this selective process. Some, like those of Warwick or Kaiser, are relatively contained in their scope, focusing on the local ergonomics of tool use inside particular academic subdisciplines, or on the influence of certain pedagogical institutions (e.g., the Cambridge mathematical tripos). But others have connected tool choice with macro-level political and cultural histories. To mention but one classic example, Theodore Porter's *Trust in Numbers* (1995) suggests that the rule-bound language of number and quantity tends to be embraced by politically accountable groups that seek credibility by appealing to an ideal of mechanical objectivity (rather than their professional expertise, for example). Quantitative evidence thus gains the upper hand in certain political environments.[11] Or to give another example, historians of the Cold War human sciences have long noted the significance of cybernetic metaphors in the modeling idiom of cognitive science, and have suggested that tools like digital computers and servomechanisms, forged in the context of military patronage, likewise got transmuted into models or theories of the human mind. However, the recent work of Jamie Cohen-Cole has broadened this story, suggesting that a number of key cognitive psychologists embraced computers as investigative tools and as models of human cognition in the context of a broader agenda to study human creativity, intellectual flexibility, and other features of the "open mind"—a mind that practiced the psychologists' own

---

Knowledge Production, Liberal Democracy, and Human Nature (New York, 2012), pp. 79–95, esp. p. 81.

10 See e.g., "World in a Matrix," in Paul Erickson et al., *How Reason Almost Lost Its Mind.*

11 Theodore Porter, *Trust in Numbers: The Pursuit of Objectivity in Science and Public Life* (Princeton, NJ, 1995).

liberal-democratic virtues. It is possible, then, to write histories that connect tool choice and theory choice with sweeping cultural and political histories.[12]

The story of the Ford Foundation's Committee on Policy and Program demonstrates the forces at work in the emergence of the Foundation's interest in "[t]he scientific study of values." The Committee was, of course, far from the only group interested in the study of values in the immediate post-war era. Even as it completed its deliberations in 1949, "academics debated the topic of values in a series of seminars on values sponsored by the Social Science Research Council at Chicago, Harvard, and Cornell Universities."[13] While the participants in the seminars ultimately concluded that "the definition of values would provide intractable problems" and cut short their study, a major research project on values got underway at Harvard: the famed interdisciplinary "Comparative Study of Values in Five Cultures" under the direction of anthropologist Clyde Kluckhohn.[14] What is particularly striking in this context is the way in which the debate inside the Committee, on how to measure and articulate "the general welfare," launched a search for appropriate practices and techniques that might help accomplish this task. As we shall see, the Committee's choices reflected a mixture of ideological, practical, and intellectual concerns: the desire by committee members for a certain objectivity and disinterestedness; the need to tame the exceedingly difficult problems facing the committee as they sought to manage their own differences of opinion; an implicit commitment to a liberal model of humans as valuing individuals and reasonable decision makers; and finally, a certain suspicion of individuals' stated expressions of value.

Valuing Research and Researching Values: The Ford Foundation's Committee on Policy and Program

By the end of 1949, the Ford Foundation's officials would announce the Foundation's funding agenda in a "Report of the Study for the Ford Foundation on Policy and Program," written by a study committee of high-powered academics and administrators. A number of the Report's proposed funding areas were targeted directly at major problems of the day: "The Establishment of

12 Jamie Cohen-Cole, *The Open Mind: Cold War Politics and the Sciences of Human Nature* (Chicago, 2014).

13 Willow Roberts Powers, "The Harvard Study of Values: Mirror for Postwar Anthropology," *Journal of the History of the Behavioral Sciences* 36 (2000), 15–29.

14 Ibid., p. 19.

Peace" in an age torn by war and Cold War, and "The Strengthening of the Economy." But a final area—"Individual Behavior and Human Relations"—was more basic in its orientation, proposing scientific research into a range of topics that, the report's authors suggested, were foundational to the other areas: "the process of learning," "the processes of communication," "group organization, administration, and leadership," "causes of personal maladjustment," and finally, "[t]he scientific study of values which affect the conduct of individuals, including man's beliefs, needs, emotional attitudes, and other motivating forces; the origins, interactions, and consequences of such values; and the methods by which this knowledge may be used by the individual for insight and rational conduct."[15]

The relatively modest placement of this bullet point in the report belies its significance in the thinking of the Foundation's study committee. Indeed, in a curious way, one root of the Ford Foundation's interest in the problem of articulating and measuring values runs back to the wording of Henry Ford's original bequest. As Henry Ford II would write to his personal legal advisor and confidant, the San Francisco lawyer H. Rowan Gaither, in the fall of 1948, roughly a year before the Foundation would first publically articulate its philanthropic agenda, "The Foundation was established for the general purpose of advancing the national welfare, but the manner of realizing this objective was left to the trustees." Naturally, the trustees sought to outsource the bulk of this task to a panel of experts, which Ford hoped Gaither would convene, "to take stock of our existing knowledge, institutions, and techniques in order to locate the areas where the problems are most important and where additional efforts towards their solution are most needed." With Ford Sr.'s estate settled and pressure building for the Foundation to start acting like a nonprofit organization, Ford hoped to begin an active program by the beginning of 1950.[16]

Gaither's presence on the committee is significant for several reasons. Apart from his close relationship with the Ford family, he was already a key member of the science-administrative elite that had formed during World War II, having served as assistant director of the famed 'Rad Lab' at the Massachusetts Institute of Technology. Subsequently, even as he was assisting Ford in setting the agenda for the new Foundation, he played a key role in the establishment

15 Ford Foundation, *Report of the Study for the Ford Foundation on Policy and Program* (Detroit, MI, 1949), p. 90.

16 Henry Ford to Rowan Gaither, 22 November 1948, Rockefeller Archive Center, New York, Ford Foundation records (hereafter RAC, FF), group 24, series 1, box 2, folder 19: "The Study of the Ford Foundation on Policy and Program Activity Reports, Oct 5 1948 to Feb 19 1949."

of the RAND Corporation, which originated in a major Air Force contract with Douglas Aircraft Company for a broad-based study of intercontinental warfare in all its complexity. Thus RAND and the early Ford Foundation would be closely joined throughout the late 1940s and early 1950s, with Gaither serving on RAND's board of trustees and a number of RAND-affiliated scholars emerging as key advisors to the fledgling Foundation.

Gaither naturally drew upon his connections in forming his committee of all-star academics—many of them with experience in wartime science administration—to develop Ford's program of giving. By December of 1948, he had assembled an all-star team for his advisory committee. In addition to himself, these included a University of Chicago political scientist; the former director of the Harvard Business School; Donald Marquis, the chair of the psychology department at the University of Michigan, recent president of the American Psychological Association, and perennial government advisor on military applications of psychological research; the New York state commissioner of education and president of the state university system; a major medical foundation director and professor at Harvard Medical School; and a Caltech physics professor who also served as advisor to the Office of Naval Research. The members directed 'divisions' of the study committee, respectively, on political science, business, social science, education, medicine, and natural sciences.[17]

Despite the study committee's members' apparently disparate interests, by January of 1949 Gaither could report to the board that "[t]here is a strong and virtually universal feeling that the profitable place for a new foundation to work is in the social sciences." Gaither stressed, however, that this did not refer to particular disciplines, but more generally to "areas of problems dealing with the relationships of man to himself, to his environment and to his neighbor and coworker. Defined thusly, the social sciences are in the middle of all disciplines, from business to health."[18] The members spent much of January and February of 1949 on fact-gathering missions, and by the study committee's third round of meetings in March they were able to identify specific research

17 On the machinations behind the establishment of the foundation and appointment of the study committee, see Francis X. Sutton, "The Ford Foundation: The Early Years," *Daedalus* 116 (1987), 41–91; and Kathleen D. McCarthy, "From Cold War to Cultural Development: The International Cultural Activities of the Ford Foundation 1950–1980," *Daedalus* 116 (1987), 93–117.

18 "Notes for Discussions with Trustees," 14 January 1949, RAC, FF, group 24, series 1, box 2, folder 19.

topics in the social sciences that needed special attention. In Gaither's progress report to the Board, he argued that "[m]any social problems (industrial relations, racial and minority tensions, political strife, for example) cannot be mitigated or solved until there is a better understanding of the critical factors in social relations. These include human values, human motivations, human organization and administration, and communication between individuals. There is great need for foundation support of activities intended to acquire more knowledge of these factors and to apply intelligently and effectively the knowledge thus secured."[19]

Suggestively, the emergence of "human values" and "human motivations" among the Foundation's interests paralleled the study committee's attempts to operationalize their original, highly vague institutional mandate of identifying programs for "advancing the national welfare." In late April, as the committee increasingly tried to pare down the enormous list of possible projects turned up in the course of its members' fact-finding missions, its leadership and staff composed and circulated a remarkable memorandum (prepared in consultation with Berkeley philosopher George P. Adams) that tackled this problem directly. Titled "What is Human Welfare? Or, How do you Tell Whether One Problem is More or Less Important than Another," the memo focused on a fundamental problem: "How do you know what human beings need, want, and desire? In other words, how do you know what human values are?" The most obvious approach to answering this question would be to listen to what people say, and indeed, "our whole democratic theory of government rests on the assumption that a human being can know what he needs and what is best for him and can consciously make that decision through representatives of their own choosing."[20] Yet while the memo ultimately came down on the side of putting "considerable emphasis upon the expressed statement of individuals as to what [...] constitutes the human values which we seek to achieve in human welfare," perhaps by an extension of "Gallup polls," it nevertheless noted that "there are some things which individuals need which they would not be able

19 "Activity Report #2 (February and March, 1949)," RAC, FF, group 24, series 1, box 2, folder 20: "The Study of the Ford Foundation on Policy and Program Activity Reports, Feb 1949 to Sep 29, 1950."

20 "What is Human Welfare? Or, How do you Tell Whether One Problem is More or Less Important than another," 22 April 1949, pp. 2–3, RAC, FF, group 24, series 1, box 3, folder 24: "The Study of the Ford Foundation on Policy and Program Planning [1948] to May 12, 1949."

to tell you themselves they need"—for example, things that lie beyond their personal experience, or novel technologies that have not yet been created.[21]

Despite the study committee's concern with "substitut[ing] the judgment of a very small group of people as to what a very large number of people need and desire," in practice, the committee's approach to identifying values remained highly technocratic, dependent on the calculations or judgments of experts for measuring the contribution of different programs to the "general welfare." A series of internal memos dating from this stage of deliberations reflects the committee's attempts to impose order on the evaluations of its own, disciplinarily diverse membership, as well as its collective decision-making processes. One memorandum exhibited a possible template for rank-ordering 'problems' for the foundation to tackle: "War and International Problems," the "Economic Cycle" and "depressions," "personal adjustment" and "mental illness," or "atomic energy." Each problem was to be assigned a "Beaufort severity rating," a "number of people affected," and a "time affected" to each one, before calculating an "aggregate rating" that was a product of all three, and that would presumably generate a ranking of the significance of the various problems. It is not clear whether the committee ever fully implemented the template system and used it in setting the foundation's priorities. However, another ranking scheme for potential "program areas," involving cumulating individual study committee members' ratings for the "importance" and "feasibility" of these areas, does seem to have been carried out.[22] In both cases, standardized paper forms and notational conventions—practical tools for reasoning and reckoning about values—provided a basic framework for structuring the committee's decision-making process.

The study of values retained a place in the final presentation of Gaither et al. to the trustees in May of the same year, which drew heavily upon the report of the Social Science Division that, in turn, closely reflected the views of Donald Marquis.[23] The Division's report argued that many of the "critical problems of human welfare" that the foundation sought to solve were essentially

21 Ibid., p. 5.

22 "Program" in RAC, FF, group 24, series 1, box 3, folder 25: "The Study for the Ford Foundation on Policy and Program: Human Welfare and the Relative Importance of Problems and Programs, 1948–1949? to Jan 6, 1950."

23 In particular, the report has some similarities to a speech Marquis gave to the Symposium on Research Frontiers in Human Relations on 6 February 1948; see Marquis, "Scientific Methodology in Human Relations," *Proceedings of the American Philosophical Society* 92 (1948), 412–3.

problems of two factors: "human attitudes and behavior" and "the organization of human effort." The report therefore called for investigation first into "human values and motives" and then into "modification of human behavior" "communications," "interpersonal influences," "personal adjustment" and "utilization of social sciences knowledge."[24] However, this report, very much like the subsequent report of the study committee, was far more skeptical of the reliability of human subjects' self-presentation as a guide to their beliefs and values. "The effective functioning of a democratic form of society is based on the *assumption that the institutions and national policies of that society reflect the true desires of the people*," the report asserted. "This is a tenuous assumption because: 1. Efforts are necessarily based on limited information; 2. It is subject to the inevitable difficulties of personal bias."[25] With few exceptions, "[p]resent attempts to gauge public opinion by quantitative surveys have been restricted to superficial and topical opinions."[26] Far greater theoretical and methodological rigor would be needed, especially the construction of "adequate measures of human motives and values," an understanding of the "psychological, cultural, and social forces which determine these basic attitudes," and comprehensive surveys of basic attitudes throughout the American population.[27]

Although the "scientific study of values" and "behavioral science" more broadly were mentioned in the 1949 report on policy and program, it would be some time before the Foundation could begin to actively implement a program of research in these areas. In the brief period between the Soviet Union's first nuclear weapons tests in August 1949 and the outbreak of a 'hot war' in Korea the following summer, program areas like "The Establishment of Peace" naturally held the spotlight—they were most immediately relevant to the problems of the age. At least initially, the Foundation's support for research on "Individual Behavior and Human Relations" primarily came in the form of large block grants to universities across the country to support self-studies on the question of how to strengthen the behavioral sciences at these institutions. The parameters for a more active program of support for research and training

24 "Summary of the Report of the Social Science Division of the Study Group: Summary of the Nine Conclusions," in Ford Foundation Archives, Group 24, Box 3, Series 1, Folder 24, "The Study of the Ford Foundation on Policy and Program Planning [1948] to May 12, 1949."

25 Ibid., p. 6.

26 Ibid., p. 7.

27 Ibid., p. 6.

in the behavioral sciences were spelled out only in the summer of 1951 thanks to the initiative of Rowan Gaither himself, in consultation with study committee member Donald Marquis and RAND Corporation political scientist Hans Speier, among others. The resulting "Behavioral Sciences Program," approved by the Foundation's board of trustees in early 1952 and supported by Gaither in his role as president of the Foundation from 1953–1956, would make grants totaling roughly $24 million before its discontinuation in 1957.[28]

Legacies for the Post-war Human Sciences

It is impossible to assess here the full impact of the Foundation's influence on the post-war behavioral sciences, much less the longer-term significance of its interest in the "scientific study of values."[29] However, it is possible to point to some specific research funded under this rubric that exemplifies the back-and-forth between theory and 'theoretical tools' suggested above. One outstanding example is the "University of Michigan Summer Seminar on Decision Processes," which was held in the summer of 1952 in Santa Monica, and had grown from a proposal to the foundation by University of Michigan psychologist Clyde Coombs, mathematician Robert Thrall, and economist Lawrence Klein. While the summer seminar left a published paper trail in the form of a rather enigmatic book, *Decision Processes*, it is not commonly remembered that this trio's initial application to the Foundation was described as a "Proposal for a Research Program in the Measurement of Values," and directly targeted the relevant bullet point in the Study Committee report.[30]

The connection between the Summer Seminar and the Ford Foundation's interest in the "scientific study of values" deepens our understanding of the choice of Santa Monica as the Seminar's venue. One obvious reason for this location was that many of that Seminar's contributors—John Nash, Oskar

28 Bernard Berelson, "The Ford Foundation Behavioral Sciences Program Final Report, 1951–1957" (Ford Foundation Archives Reports Collection, accession number 76599), p. 2.

29 For a survey of the extent of the foundation's program, see especially Mark Solovey, *Shaky Foundations: The Politics-Patronage-Social Science Nexus in Cold War America* (New Brunswick, NJ, 2013), chapter 3.

30 Robert M. Thrall, Clyde H. Coombs and Robert L. Davis (eds.), *Decision Processes* (New York, 1954). The initial grant application is "Proposal for a Research Program in the Measurement of Values" (Ford Foundation Grant File #52–98 [microfilm]). For more discussion of the context behind the Summer Seminar, see Paul Erickson, *The World the Game Theorists Made* (Chicago, Forthcoming), chapter 4.

Morgenstern, and Alan Newell, to name only a few—were affiliated with the nearby RAND Corporation in various capacities. But it is also worth noting that something like the "scientific study of values" had been a central part of RAND's mission from the start, just as it had for the Ford Foundation. As Martin Collins has traced so thoroughly, RAND's guiding objective in its early years was the assessment of 'military worth,' that is to say, the quantification of the benefit conferred on military operations by various choices of technology, tactics, and personnel. As RAND's ambitions expanded from the evaluation of individual technologies to programming the Air Force budget, and ultimately, to sizing up the Cold War international system as a whole, RAND dramatically expanded the disciplinary diversity of its experts, cultivating connections with economists, political scientists, anthropologists, and sociologists, among others.[31] In the face of this buzzing confusion of disciplines, the theoretical framework of military worth analysis, drawing especially on ideas from game theory and utility theory, held out hope that these experts' assessments of the various technologies, tactics, and strategies available for use in the dawning Cold War could be expressed in a common metric of 'worth' which would serve as a guide for action.

The connection between the Summer Seminar, RAND, and the Ford Foundation's interest in 'values' also helps in making sense of the peculiar blend of contributions that ultimately found their way into *Decision Processes*. On the one hand, the volume contained a number of purely logical investigations of utility theory, preference orderings, and decision-making under conditions of uncertainty. These built on some of the most celebrated theoretical publications in the post-war human sciences, especially John von Neumann and Oskar Morgenstern's *Theory of Games and Economic Behavior* (which introduced a rigorous axiomatic formulation of a numerical utility function) and Kenneth Arrow's *Social Choice and Individual Values* (which called attention to the problems involved in aggregating the preference orderings of individuals to produce a collective preference ordering).[32] Other studies discussed in the volume were experimental in approach, 'testing' various theoretical predictions against reality—for example, in reports of experiments on learning theory, or a study of "Some Experimental n-Person Games," comparing

31 Martin J. Collins, *Cold War Laboratory: RAND, the Air Force, and the American State, 1945–1950* (Washington D.C., 2002), especially chapter 4: "Reshaping RAND: Air Warfare as a Domain of Research."

32 John von Neumann and Oskar Morgenstern, *Theory of Games and Economic Behavior* (Princeton, NJ, 1944); Kenneth Arrow, *Social Choice and Individual Values* (New York, 1951).

predictions and experimental data.[33] But a final category of studies turned to the theoretical apparatus of utility measurement and preference orderings as inspiration for new experiments that would help behavioral scientists gain insight into the beliefs, attitudes, and values of their human subjects. Thus Clyde Coombs reported on a study in which groups of undergraduates at the University of Michigan would try to arrive at a collective rank-ordering of the aesthetic attractiveness of isosceles right triangles.[34] Likewise, another study gave Stanford undergraduates modified IQ tests in which the students could gain additional points by colluding at the expense of a "ringer" who appeared to race through the test and finish well in advance of the others. The essential question: in a game situation, what do subjects value more, absolute gains or gains relative to their peers?[35]

Hunter Heyck has recently commented on the proceedings of the Seminar, suggesting that they reflected the post-war emergence of new and general "sciences of choice," freed from the assumption that the things doing the choosing are individual humans.[36] Certainly, the blend of utility theory, learning theory, social choice theory, and game theory represented in the proceedings provided a new language sufficiently flexible to encompass the behavior of humans, lab rats, committees, and potentially entire societies. But the mathematics of 'decision processes' was not merely a language, nor did it just provide a 'model' of behavior: it simultaneously suggested a collection of procedures and practices for the Seminar's social and behavioral scientists themselves. Just as the tables and rank orderings of the Study Committee constituted tools for eliciting, articulating, and aggregating the 'values' of the committee members, a set of methodological precepts for measuring and aggregating values was implicit in the new 'theories' of decision-making circulating in that summer of 1952—precepts that were clinical, mechanical, and, to some extent, skeptical of the representations made by human subjects, whether as individuals or collectively. Secure knowledge of the values held by other individuals, societies, or cultures did not result from extensive experience or from a certain kind of trust or closeness between observer and observed, but ideally, from adherence to rigorous techniques for measuring and aggregating.

33 Gerhard K. Kalisch, J.W. Milnor, J.F. Nash and E.D. Nering, "Some Experimental N-Person Games," in Thrall, Coombs, Davis (eds.), *Decision Processes*, pp. 301–27.

34 Clyde H. Coombs, "Social Choice and Strength of Preference," in Thrall, Coombs, Davis (eds.), *Decision Processes*, pp. 69–86.

35 Paul Hoffman, Leon Festinger and Douglas Lawrence, "Tendencies Toward Group Comparability in Competitive Bargaining," *Human Relations* 7 (1954), 231–253.

36 Heyck, "Producing Reason."

Yet as the more mathematical results of the Summer Seminar seemed to suggest, there was a limit to how far these techniques could take experimenters toward understanding their subjects. The determination of a subject's beliefs and prior probabilities, like values, could not be placed on secure theoretical foundations—nor could any kind of aggregative scheme for values that the Seminar's contributors could invent, as Arrow's impossibility theorem suggested.[37] The efficacy of this particular style of inquiry into beliefs and values thus lay on shaky logical ground, even as it retained its popularity.

For a time, the popularity of this style of inquiry ran strong in certain quarters. Inasmuch as they were intended to shape the setting of policy, techniques like military worth assessment, 'systems analysis,' or 'mathematical programming,' depended on the articulation of coherent measures of values. Such techniques, initially forged to assess war-fighting technologies and tactics, subsequently filtered into the assessment of urban, welfare, and health care policy, as defense think tanks sought to diversify their contract base, and the social programs of the 1960s provided them with alternative revenue streams.[38] Techniques and concepts from the policy sciences, especially game theory, utility theory, and social choice theory, invaded even the most obscure precincts of philosophy and ethics, cropping up not only in Arrow's *Social Choice and Individual Values*, but in debates over John Rawls' famous thought experiment in which social values could be assessed by placing subjects behind a "veil of ignorance," which hid their eventual position in a would-be new society from them, while they were asked to distribute rights, positions and resources.[39]

37 See e.g., Leo A. Goodman, "On Methods of Amalgamation," pp. 39–84; Clyde H. Coombs and D. Beardslee, "On Decision Making Under Uncertainty," pp. 255–85; Stefan Vail, "Alternative Calculi of Subjective Probabilities," pp. 87–98; and Herbert G. Bohnert, "The Logical Structure of the Utility Concept," pp. 221–230 in Thrall, Coombs, Davis (eds.), *Decision Processes.*

38 See especially Jennifer S. Light, *From Warfare to Welfare: Defense Intellectuals and Urban Problems in Cold War America* (Baltimore, MD, 2003); David Jardini, "Out of the Blue Yonder: The Transfer of Systems Thinking From the Pentagon to the Great Society, 1961–1965," in Agatha C. Hughes and Thomas P. Hughes (eds.), *Systems, Experts, and Computers: The Systems Approach in Management and Engineering, World War II and After* (Cambridge, MA, 2000); David Jardini, *Out of the Blue Yonder: The RAND Corporation's Diversification into Social Welfare Research, 1946–1968* (Ph.D. dissertation, Carnegie Mellon University, 1996).

39 John Rawls, *A Theory of Justice* (Cambridge, MA, 1971). On casting the arguments of Rawls in a game-theoretic idiom, see John C. Harsanyi, "Can the Maximin Principle Serve As a Basis for Morality? A Critique of John Rawls's Theory," *The American Political Science Review* 69 (1975), 594–606.

Yet America in the 1960s witnessed a reaction to the technocratic optimism of the age, as humans proved resistant to having their values assessed and neatly aggregated by following the kind of rules and procedures that the Ford Foundation's study committee had found so attractive. An expert committee, a military chain of command, or a corporation: these were the kinds of organizations for whom this style of value-assessment might just possibly work. It applied less neatly, perhaps, to the raw clash of ideals and wills that characterized America's "age of fracture."

Index of Names

Aarons, J. 54, 55
Adams, G.P. 281
Adams, R. 15, 16
Adenauer, K.H.J. 175
Ahlmann, H.W. 21
Amadae, S.M. 275
Amstrup, N. 270
Andersen, E. 268
Anderson, C.P. 232–236
Anderson, W. 207
Arisz, W.H. 173
Arnold, R.T. 24
Arrow, K.J. 275, 285, 287
Ashford, B.K. 192
Atwood, W.W. 48
Ayrout, H.H. 210

Baneke, D. 6
Barendsen, G.W. 158
Baunsgaard, H. 268
Baylis, J. 62, 63
Beenakker, J.J.M. 158
Beets, W. 89
Bennett, E.F. 30
Bergh, S.J. van den 108, 115
Berkner, L.V. 22–24, 27, 28, 31, 38
Birn, A.-E. 196
Bjerknes, V. 219
Bleeker, R.D. 160
Bloembergen, N. 153, 157, 173
Boel, N. 268
Boer, J.H. de 105–107, 109, 113, 124, 126
Boettcher, A. 87, 88, 90, 98
Bohr, N. 150–154, 157
Bolin, B. 223
Bolinger, R.E. 148
Bolkestein, F. 156
Boone, W.W. 149
Bosch, H. 159
Boyd, G.E. 147
Braams, R. 158
Brabers, M.J. 160
Brode, W.R. 14, 18, 19, 28–31, 35
Bronk, D. 27, 28, 30
Brosio, M. 39
Brown, T.M. 150
Bundy, M. 64
Burgers, W.W. 157
Burr, W. 64
Bush, V. 15, 16, 18, 48, 49, 110, 152, 154
Byrnes, J.F. 21, 251

Callahan, A.L. 182
Cals, J.M.L.T. 171
Carr, W.J. 15
Carr-Saunders, A.M. 209
Carrara, N. 54, 55
Case, F.H. 234, 236
Casimir, H.B.G. 170, 172
Cath, P.G. 160
Chang, K. 207
Charney, J. 222
Cmentek, A. 123
Cohen, H. 149
Cohen, J.A. 113, 114, 125, 132
Cohen-Cole, J. 277
Collett, J. 178
Collins, M. 285
Conant, J.B. 17, 24
Condon, E. 110
Coombs, C. 284, 286
Coster, D. 106
Coster, H. 159
Cray, S. 67
Critchfield, R. 210
Cronkite, W.L. 264
Cullather, N. 202

Davis, F.B. 148
Delft, D. van, 2
Delsman, A. 159
Doel, R. 2, 58
Dongen, J. van 2, 3
Dresser, C.S. 245
Dubbeldam, P.S. 158
Dullemond, C. 159
Dulles, J.F. 25–29

Einstein, A. 42, 43
Eisenhower, D.D. 25, 26, 30–32, 41, 49, 50, 59–61, 64, 85, 117, 150–152, 235
Ekbladh, D. 200
Eliassen, A. 223
Eliot, C.W. 192
Emerson, H. 207
Engerman, D. 138
Erickson, P. 4
Eriks, K. 158
Espinoza, M. 205

Feestehausen, H.H. 150
Fermi, E. 44
Finlay, C.J. 206
Fjørtoft, R. 223
Fock, C. 90
Ford, G.R. 68
Ford, H. 274, 279, 280
Ford, H. II, 279
Forman, P. 38, 44, 169
Franco Bahamonde, F. 167
Freedman, R. 149
Fuchs, K. 61
Fulbright, W.J. 136

Gaither, H.R. 279–282, 284
Galley, R. 65, 69–71
Gammell, J. 154
Gates, F.W. 192
Gaulle, C. de 64, 70, 71
Geurts, J. 149
Gigerenzer, G. 276
Giscard d'Estaing, V.M.R. 64
Gold, H. 149
Gorter, C.J. 153, 173
Graham, P.L. 35
Greulich, W.W. 24
Groneveld, E.W. 160
Groot, A.D. de 148
Groth, W. 82, 96, 97

Haaijer, G. 160
Hansen, H.C. 247, 264
Harper, K. 5
Harsanyi, J.C. 275
Hartling, P. 268
Hartog, H. den 153, 158
Hassig, H.J. 159
Haynes, C.V. 253
Hedtoft Hansen, H. 256
Heer, J. de 159
Heertjes, P.M. 160
Heffener, H. 118
Heiser, V.G. 207
Herman, E. 275
Hersch, S.M. 240
Hertz, H.G 80
Heyck, H. 72, 275, 286
Heymann, M. 5
Hilgevoord, J. 153, 159
Hillenkoetter, R. 19
Hobsbawn, E.J.E. 174
Hoeneveld, F. 3
Hollander, A.N.J. den 143
Holstege, C.J.M. 159
Hoover, J.E. 15–17
Hooyman, G.J. 158
Hornig, D.F. 238–240
Houben, G.M.M. 130
Hugenholtz, N.M. 153, 154, 158
Huizenga, J.R. 154
Huxley, J.S. 208

Igo, S.E. 275
Isaac, J. 276
Ismay, H.L. 167, 168
Iyer, S. 208, 209

Jackson, H.M. 41
Jandrey, F. 62
Janowitz, M. 149
Jensenius, H. 7
Jobert, M. 66
Johnson, L.B. 64, 238–240
Joliot-Curie, F. 24
Judt, T.R. 167

Kaiser, D. 2, 72, 169, 276, 277
Kai-shek, C. 27
Kamerlingh Onnes, H. 151
Kármán, T. von 40
Kauffmann, H. 251
Kennedy, J.F. 31, 64
Kenny, G. 232

Kerry, J.F. 136
Khan, A.Q. 4
Killian, J. 30, 41, 110, 152, 154
Kissinger, H.A. 60, 63, 65–70, 73
Kistemaker, J. 3, 4, 77, 79–82, 87–100, 117, 153, 158
Kistiakowsky, G.B. 30, 31
Klein, C. 189
Klein, L. 284
Kluckhohn, C.K.M. 278
Kluyver, J.C. 153, 158
Knudsen, H. 5
Koch, H.H. 263
Koepfli, J. 41
Kolstadt, G.A. 82, 88
Komer, R.W. 212
Korringa, J. 159
Krag, J.O. 247, 248, 263, 270, 271
Kramer, P. 138
Kramers, H.A. 106, 152
Kramers, H.C. 158
Krige, J. 3, 4, 78, 114, 146, 168, 173, 182, 212
Krijgsman, C. 160
Kristensen, T. 256
Kronberger, H. 98
Krusemeyer, H. 159
Kruyt, H.R. 107, 109
Kuc, J. 149
Kuiper, G. 5, 33, 34, 172, 173
Kuster, C. 7

Laird, M.R. 240
Landsberg, H. 233
Lange, H. 41
Langmuir, I. 227, 228, 231, 241
Laursen, V. 249, 250, 254, 256
Lebovic, S. 142
Lewis, A. 209
Lied, F. 53
Lilienthal, D.E. 200
Lith de Jeude, O.C.A. van 106
Lorentz, H.A. 151
Los, J. 97
Lubbers, R.F.M. 153
Ludwig, A.M. 178
Luning Prak, N. 159
Lunteren, F. van 2
Lysenko, T. 12, 19

Maas, A. 2
Macauley, J.B. 51, 53
Machle, W. 19
Macmillan, H. 64
Maharaj, A. 182
Malaurie, J. 257
Maris, W.H. 110
Marquis, D.G. 280, 282, 284
Marr, D. 211
Marshall, C.L. 84
Marshall, G.C. 21, 208
Martino, G. 41
Marvel, J. 254
Maverick, M. 199
Mazower, M. 199
McCoy, A.W. 211
McLaughlin, A.J. 207
McNamara, R.S. 248, 262
McVety, A. 203
Meer, S.M.L. van der 159
Meijer, P.H. 158
Meijer, S. 89
Merton, R. 17
Messmer, P. 71
Metzger, H.B. 149
Meyers, H. 83
Meynen, J. 115
Miller, P.G.E. 144
Mirowski, P. 275
Mitchell, T. 210
Mølgaard, J. 245
Morgenstern, O. 285

Nash, J.F. 275, 284
Nasser, G.A. 61
Nauta Lemke, H.R. van 154, 155, 160, 161
Needell, A. 37
Neumann, J. von 221–223, 227, 275, 285
Newell, A. 285
Nielsen, H. 5
Nielsen, K.H. 5
Nijboer, B.R. 157
Nixon, R.M. 60, 63, 65, 69, 240
Norman, A.C. 268
Norstad, L. 40, 43, 45
Nussbaum, R.H. 161

Odum, H.W. 200
Okkerse, B. 160
Oldendow, K. 249, 250
Olesen, J. 267
Olesen, T.B. 271
Ooms, A.J.J. 125, 134
Oort, J. 171, 173
Oppenheimer, J.R. 34 , 136, 151
Ormondt, J. van 5, 105, 106, 109, 113, 122, 124–135
Orville, H.T. 235
Oswald, J.W. 148, 149
Overbeek, J.T.G. 173
Overstrate Kruysse, M.P.C. van 160

Packard, R.M. 207
Pais, A. 151
Palmer, S. 196
Paris, C.H. 158
Person, L. 41
Pestre, D. 173
Petersen, H. 249, 250
Petersen, V. 262
Phillips, N.A. 225
Piket, J. 111
Plummer, B.G. 204
Pompidou, G. 64
Popkin, S.L. 210
Porter, R. 175
Porter, T.M. 277
Poulis, N. 158
Prins, J.E. 161
Purcell, E.M. 153

Rabi, I.I. 42–52, 56, 57
Raj, K. 72
Ramsey, N.F. 42, 45–53, 56, 57
Rasmussen, G. 251
Rawls, J. 287
Reerink, H. 161
Reichelderfer, F.W. 221
Reinke, N. 175
Reza Pahlavi, M. 202
Rietjens, L. 153, 158
Riley, J.C. 198
Ring, N.J. 191
Rinnooy Kan, A.H.G. 156
Robertson, H.P. 42–53, 56, 57
Rockefeller, J.D. 191
Roosevelt, F.D. 17, 25, 199, 218
Rose, W. 192, 194
Rosenberg, E. 26
Rosenberg, J. 26
Ross, E.A. 209
Rossby, C.-G. 223, 225
Rostow, W.W. 238–240
Rupp, J.C.C. 138, 140

Sauvy, A. 209
Schaefer, V. 228
Schapiro, L. 207
Schermerhorn, W. 5, 106, 107, 116, 120, 172
Schlesinger, A.M. 144
Schlesinger, J.R. 70
Schliekelman, R.J. 159
Schulman, B. 199, 200
Scott, J.C. 210
Sebesta, L. 176
Secord, J. 34, 60, 72
Seitz, F. 53
Selassie, H. 203
Serup, A. 260
Shapin, S. 72
Simony, C.F. 253
Simpson, B.R. 202
Sizoo, G.J. 5, 103–111, 114–120
Sleebos, F. 125, 126, 129, 130
Slotboom, H.W. 172
Smith, J.M. 147
Smith, R.W. 174
Snowcroft, B. 70
Soest, J.L. van 105, 107
Solovey, M. 187
Sørensen, M. 254
Staf, C. 116, 119
Stalin, J. 26, 31
Steacie, E.W.R. 47–50, 52, 57
Steelman, J.R. 21
Strath, W. 60
Strauser, W.A. 83
Streefland, A. 3
Stumpers, F.L. 158
Suri, J. 69

Teller, E. 30
Terlouw, J.C. 156
Thijsse, J.Th. 160
Thrall, R. 284
Tichelaar, G.W. 158
Tilley, H. 199
Timmermann, A. 7
Tinbergen, N. 173
Tizard, H. 109
Trachtenberg, M. 39
Trier, A.A. van 160
Troost, L. 160
Truman, H.S. 21–26, 150, 201
Tse-tung, M. 27, 63
Turchetti, S. 2

Ullman, R. 63–65, 68

Vandenberg, H. 18
Vassy, E. 55
Veltman, B.P.T. 158
Verrijn Stuart, A.A. 152, 157
Verschuren, J.P. 160
Visser, S.H. 96
Vonnegut, B. 228
Vries, H. de 157
Vries, J.P. de 160

Waals, J.D. van der 151
Wageningen, R. 159
Walker, A.G. 42
Walske, C. 246
Wang, J. 6
Warwick, A. 276, 277
Waterman, A.T. 110
Wells, A.A. 82–85
Williams, W.A. 212
Windmuller, J.P. 150
Woltjer, H.R. 90
Woods, R. 137
Wouthuysen, S. 151

Zahir Shah, M. 202
Zippe, G. 81, 82
Zuckerman, S. 45, 46, 52, 57
Zworykin, V. 221, 227

www.ingramcontent.com/pod-product-compliance
Lightning Source LLC
Chambersburg PA
CBHW071249180926
51613CB00002B/181